The Media Mindset

How B2B Brands Win By Thinking Like Media Companies

Jim Gilbride | Scott Anthony | Eric Sandy

THE
MEDIA
MINDSET

HOW B2B BRANDS WIN BY
THINKING LIKE MEDIA COMPANIES

JIM GILBRIDE | SCOTT ANTHONY | ERIC SANDY

ISBN: 979-8-9917616-9-7

TILT
PUBLISHING

Tilt Publishing
700 Park Offices Drive, Suite 250
Research Triangle, NC 27709

Table of Contents

Authors' Note:

Where We Came From and Where We're Going

Before we built Encore360, we spent more than 35 years combined in traditional B2B publishing. We launched and scaled media brands that dominated their markets because we understood one thing better than anyone:

The list *is* the business.

The audience is the product. And no real media company will ever fully sell it to you.

They might rent you a piece of it. They might sell you a sponsored segment. But they will never hand over the full list, because if they did, you'd never need them again.

That's the truth behind most advertising deals:

- You don't get the whole subscriber list.
- You don't get full email access.

- You don't get the full attendee data from the trade shows.
- You get just enough access to keep coming back for more.

Sound familiar?

Trade shows work the same way.

You pay five or six figures to exhibit, year after year. But you never walk away with a full list of who attended the event. If you did, you'd stop buying the booth. So the organizers guard that data fiercely. Because they know: if you own the list, you own the leverage.

We've seen this game from the inside, and we've played it.

Which is why, when we launched Encore360, we flipped the model. We help companies build what no one will ever sell them:

- An audience they own
- A media engine they control
- A brand presence that compounds in value over time

We took everything we learned from B2B publishing—about editorial cadence, format development, voice control, and brand trust—and applied it to modern content marketing.

The result?

Companies stop chasing other people's audiences. They start growing their own.

They stop buying attention. They start earning it.

They stop acting like vendors. They start operating like media brands (because, as we'll explain, your business is also a media brand, whether you want it to be or not).

Here's the thing: In 2025 and the years to come, your company has more tools, more reach, and more opportunity to build media gravity than we ever had in publishing.

If you're still renting audience access from someone else, you're playing a game you can never win.

Introduction:

Marketing Isn't What You Think It Is

L et's get something out of the way: This is not another book about marketing trends. You don't need more LinkedIn platitudes, funnel diagrams, or breathless updates on the latest algorithm shift. You need a way to *think* that helps your team create better work more consistently, without getting buried in noise or burnout.

That's what this book is about.

It's a practical playbook for small B2B marketing teams who are tired of chasing one-off campaigns and ready to build something more powerful: a content ecosystem that works like a media company.

Because here's the thing no one tells you when you're building a B2B brand in the digital age: you're not just marketing a product. You're earning attention. Sustaining it. Turning it into trust.

And if you want to earn trust at scale, you have to think like a publisher.

The Shift: From Campaigns to Ecosystems

The old model of B2B marketing was built for a different era. You ran a campaign, you collected leads, you handed them to sales. Rinse, repeat. It was a transaction-based mindset dressed up in personas and PowerPoint.

But that world is gone. Buyers today aren't waiting around for your gated white paper. They're finding answers, forming opinions, and vetting solutions long before they ever fill out a form. They don't want to be captured; they want to be served.

This shift isn't theoretical. It's operational. It changes how you plan, produce, distribute, and measure everything you do.

Media companies figured this out decades ago. They don't chase clicks; they build audiences. They don't run campaigns; they run shows. They don't speak broadly to everyone; they speak clearly and consistently to the people who care.

Think about that for a second. When a media company launches a new product, it's rarely a one-and-done splash. It's a recurring format. A weekly column. A monthly video. A podcast that earns its place in your routine. These formats create consistency, and consistency creates trust. That's how loyalty is built—not through a one-time CTA, but through a steady cadence of value.

And it's not just about quantity. It's about tone, perspective, rhythm. The media mindset recognizes that you win not by interrupting but by becoming a part of your audience's information diet. You become a trusted voice. Not because you're everywhere, but because you keep showing up where it counts with something worth saying.

That's the mindset we're after.

What This Book Is Not

This is not a book about ad spend, funnels, or viral hacks. You will not find any secrets to 10x your MQLs.

We're not here to optimize your click-through rate. We're here to help you become the publication of record for your niche.

If your job is to generate demand, grow trust, and build authority in a B2B space, then your team needs to operate more like a newsroom than a sales support department. That's not a metaphor. It's a mandate.

This book won't teach you how to go viral. It will teach you how to go deep, stay consistent, and build something people actually want to subscribe to.

What This Book *Is*

This is a field guide. A map. A challenge.

It will help you:

- Shift your team's mindset from one-off marketing to continuous content.
- Build editorial habits that lead to repeatable, scalable formats.
- Structure your workflow like a media operation.
- Align your internal experts around consistent storytelling.

- Distribute your ideas across owned and rented channels with purpose.

- Use AI and lightweight video tools without losing your voice.

- Become a brand that educates, entertains, and earns loyalty—before the pitch ever happens.

You don't need a massive team or a huge budget. You need the right mental model. This book gives you that model, with tactical steps to make it real.

Why the Media Mindset Matters Now

The internet is loud. The trust deficit is real. And your audience has a very good filter for BS.

Every day, they scroll past hundreds of headlines, posts, pitches, and emails. Their attention is rationed. Their time is expensive. And they've learned—sometimes the hard way—that most content isn't worth the click.

In that environment, marketing can't afford to be shallow. It has to be useful. It has to be repeatable. And it has to earn its spot in someone's feed, inbox, or brain.

That means content can't just be a checkbox on your to-do list. It has to be part of how your company shows up in the world—consistently, credibly, and with something valuable to say.

The companies that win today aren't the ones with the most content. They're the ones with the clearest voice and the most consistent cadence. They show up with insight. They bring receipts. They act like teachers, not vendors.

They're not trying to manufacture urgency; they're building familiarity. They're not shouting into the void; they're earning a seat at the table, over time, through relevance and rigor.

Sound familiar? That's what good media does.

And it's what your marketing can do—if you stop trying to sell and start trying to serve.

Why This Book Exists

We wrote this because we kept seeing smart teams get stuck.

They knew content was important. They had the talent and the buy-in, but they were still approaching marketing like a series of disconnected events:

- A campaign for this trade show.
- A few blogs when someone has time.
- A LinkedIn post that dies after two likes.

They weren't thinking like publishers. They weren't building systems. They weren't developing repeatable formats or aligning their internal experts around shared narratives.

They were working hard, but they weren't building anything lasting.

So we started helping teams shift how they think. And when they did, everything changed. Content stopped feeling like a chore. It became the product. It became the brand.

Not overnight. But with consistency.

This book is the distillation of that shift. It's the roadmap we wish we had ten years ago.

Who This Is For

This book is for the B2B marketing leader with a team of three and a hundred ideas.

It's for the solo marketer who has to wear nine hats and is tired of feeling behind.

It's for the founder who knows they have something to say but doesn't know where to start.

It's for the skeptical exec who thinks content is fluff but can't deny the impact of a sharp POV in the market.

It's for the teams who want to stop playing catch-up and start playing long game.

A Quick Word on Format

Each chapter is built to stand alone if that's how you read it. You can read cover to cover or jump to what you need.

We'll lead with a clear thesis, back it with real examples, and end with takeaways you can use tomorrow. The voice will stay sharp, direct, and human—because the last thing the world needs is another vague business book.

You'll see phrases repeated. That's intentional. Repetition builds rhythm. It sharpens ideas. It reinforces the shift.

Think of this book like a newsroom: clear beats, smart structure, consistent voice. That's not just how it's written. It's what we want you to build.

Ready?

Then let's begin.

Let's ditch the one-and-done. Let's stop thinking like vendors, and start thinking like showrunners.

Let's turn marketing into media.

Let's go.

Why We Built Encore360 in the First Place

Marketing is changing fast. Buyers don't wait for sales calls. They research, they read, and they form opinions long before a company ever gets the chance to pitch them. Content—strong, strategic, and consistent—is what shapes those opinions. But here's the problem: Most companies aren't executing it well.

We see this problem time and time again across all markets.

The Problem

Some businesses have great ideas but no consistency. Others throw money at digital marketing without a real strategy. Then there are the companies sitting on a goldmine of insights but doing nothing with them.

That's why we started Encore360.

"We're media people, and content production and content marketing is what we do," says Encore360 CEO Jim Gilbride. "I was a publisher in the B2B media industry and I continued to look at our advertising clients' websites and saw that they struggled to come up with consistent content for their blogs."

Content wasn't the problem. Execution was.

"There was a lot of start-and-stop or not start at all–or a lack of engaging content," Jim adds.

Meanwhile, Encore360 President Scott Anthony saw another challenge present itself over the past decade.

"We continued to see shifts in how businesses were spending marketing dollars," Scott explains. "Clients were spending dollars, but not always on the most effective things."

The Opportunity

Businesses knew they needed content. They just didn't know how to do it effectively. The disconnect was clear—and so was the opportunity.

The solution wasn't just more content—it was better content, aligned with strategy. Content that positions businesses as the go-to resource in their industry. Content that builds trust. Content that drives sales.

"The classic marketing funnel has changed. Trust is built outside of talking to a sales rep," Jim explains.

Buyers today aren't waiting for someone to sell to them. They're making decisions on their own. If your content isn't showing up when they search, you don't exist in their world.

"If ChatGPT can't crawl your website for information, you're not going to be recommended," Jim warns.

It's not just about Google rankings anymore. AI-driven discovery tools are rewriting the rules. The question isn't whether content marketing matters—it's whether your content is visible enough to be found.

Speed Matters

"You have about seven seconds to get that reader's attention and this trend is not slowing down," Scott explains.

You still need in-depth, valuable content—but if you don't hook people immediately, they'll move on.

"You have to get their attention first, or they won't stick around for the rest," he adds.

Encore360 helps businesses create content that grabs attention and delivers lasting value. Because if you're not capturing your audience, your competition is.

The Reward

"Building something from nothing and having complete creative control to create a content marketing business," Jim says, "has been one of the most rewarding aspects of this journey."

But the best part? Seeing our clients win.

"Watching our clients' businesses thrive as we have an impact on their companies," Jim says, "is what makes the work worth it."

Scott agrees. "Every day is truly different," he says.

The fast pace, the problem-solving, the ability to take a company's knowledge and turn it into something that drives revenue—that's what keeps us excited about what we do.

The Stakes

The market for content marketing isn't slowing down. With a projected value of $600 billion and an annual growth rate of 15-20%, content isn't just part of marketing—it is marketing.

"The threat becomes: are they going to read on your competition's website or are they going to read on your website?" Jim points out.

For companies that haven't invested in content, the risk isn't just falling behind—it's becoming irrelevant.

"You have to participate in it from a content marketing perspective or, again, you're going to be left behind," Jim says.

Scott puts it even more bluntly: "Adapt quickly. Communicate directly. Provide value."

That's why we started Encore360—because we knew businesses needed a content strategy that actually works. If you don't take control of your content, someone else will.

1

The Case for a Media Mindset

et's start with a simple question: What is marketing's job? If your answer is "to support sales," you're not wrong. But you're not entirely right either.

Marketing isn't just here to tee up leads for sales to close. Not anymore. Marketing's job is to build trust at scale. To shape the narrative. To make your company known, respected, and remembered *before* the first sales call even happens.

To do that, you have to think like a media company.

Traditional Marketing vs. Media-Driven Marketing

Traditional B2B marketing is campaign-based. It operates like a fireworks show: briefly dazzling, quickly forgotten. You launch a campaign, target some personas, run ads, publish a few pieces of content, collect leads, and move on to the next thing. That's not a system, it's a sugar high.

Media-driven marketing is different. It's not built around bursts of activity. It's built around rhythm. Around habits. Around the understanding that a single blog post doesn't change minds—but a weekly cadence of sharp, useful content can.

Where traditional marketing asks, "What do we want to say?" Media-driven marketing asks, "What does our audience need to hear, and how can we earn their attention over time?"

Traditional marketing creates a sequence of unconnected efforts. Media-driven marketing builds an ecosystem of interlocking formats, voices, and channels.

This isn't just semantics. It's strategy.

Take, for example, a B2B company selling water treatment systems to municipalities. The traditional approach might involve a three-month campaign: a white paper download, two webinars, and an email blast. Fine.

But a media-minded approach might launch a monthly "Behind the Flow" interview series with public works directors, a biweekly newsletter curating regulatory news, and a podcast hosted by a veteran plant operator. Each asset becomes part of a network. Each one feeds the next.

This is how you build relationship equity. Not with grand gestures, but with reliable presence.

Why Attention Is the New Currency

Let's be blunt: Nobody cares about your product until they trust your perspective. And trust starts with attention.

We are competing not just with our competitors but with everything: social feeds, Slack pings, streaming services, inbox noise. Attention is the gatekeeper. If you don't earn it, you don't get a shot.

And here's the kicker: Attention is no longer earned once. It's earned repeatedly. Every email, every post, every article is a chance to show—not tell—that you're worth the reader's time.

This is why frequency matters. This is why voice matters. This is why content can't be treated like a quarterly checklist.

Your marketing doesn't need to go viral. But it does need to show up with substance. Repeatedly.

Lessons From Media Companies

Media companies don't treat publishing as an afterthought: It's the product.

Here are three lessons B2B brands should steal shamelessly:

1. Frequency builds familiarity.

Repetition is recognition. And recognition precedes trust.

If your team is debating whether two blog posts per month is "too much," you're asking the wrong question. The right question is: *How often do we want our audience to remember us?*

Think about Axios. Their daily email formats are designed to be skimmed in three minutes or less. Readers don't just tolerate that frequency, they rely on it. What if you told readers up front just how many minutes it would take to read your weekly newsletter? And what if that helped draw them in?

Let's keep pushing: Can you publish a weekly tip on LinkedIn? A monthly field report from your service techs? A recurring five-minute audio clip from your CEO? Each format becomes a slot in your audience's mental calendar.

2. Voice builds loyalty.

The best media brands sound like some*one*, not some-*thing*. They have opinions. Humor. Personality.

Basecamp built a massive following with blunt, opinion-ated content on remote work and software design. You didn't have to agree with them—but you remembered them. And you *trusted* that they believed what they were saying.

Most B2B brands sanitize their content until it's unrecog-nizable. But a distinctive voice is a competitive advantage. Think of the most compelling people you know; you enjoy their bold voice, right? Maybe it's too much at times, but you know you're going to walk away with an interesting new idea after talking with them.

Write like a person. Take a stance. Say what others are too timid to say. That's how you stand out.

3. Trust is built in public.

Media companies don't hide their best ideas behind paywalls or sales reps. They earn influence by giving away real value. Repeatedly. Publicly.

This means your how-to guides should be genuinely helpful—not glorified brochures. Your newsletters should teach something new. Your webinars should feel more like talk shows than sales decks.

You want people to think: *If this is what they give away for free, I can only imagine what they're like to work with.*

That's the shift.

B2B Isn't Boring—It's Under-Produced

If you've ever heard a CEO say, "We're in a boring industry," here's your answer: *Then be interesting.*

Boring industries are just unstoried industries. They're full of smart people solving complex problems—and that's gold, if you know how to frame it.

Take GE's "Unimpossible Missions" video series. They turned engineering tests into cinematic stunts. Or OpenView's Build podcast, where investors interview early-stage SaaS leaders. Or Grammarly Business's LinkedIn posts: practical, nerdy, and deeply relevant to B2B writing teams.

Even wastewater treatment (yes, even that) can shine: Interview the operators. Show the control room. Compare data dashboards. Invite readers behind the curtain.

The trick isn't hype. It's *curation*. It's knowing what your audience obsesses over and packaging that with care and clarity.

Here's how to start:

- Build a recurring content format tied to your buyer's core questions.
- Use plain language. Avoid jargon unless you're owning it.
- Bring the field into the feed: show processes, people, progress.

- Layer content formats: blogs become posts become email snippets become quote cards.

Your content shouldn't feel like homework. It should feel like a shortcut to knowing what matters.

Where Your Audience Lives (Hint: Everywhere)

Today's buyers don't camp out on your website—they encounter you in fragments. A quote on LinkedIn, a snippet in an AI summary, a clip on YouTube. As executive coach and content entrepreneur Kathryn Aragon told us, "Your content is your brand, no matter where it lives."

And she's right. If your entire content strategy depends on driving people to your homepage, you're optimizing for a vanishing behavior. Today's buyer doesn't binge your blog. They catch glimpses of your thinking across platforms—over days, weeks, even months. They scroll past a headline. They see a pull-quote. They hear your voice on a podcast while making breakfast.

These aren't throwaway moments. They're trust-building moments. They're how reputations form in the real world: gradually, accidentally, and through repeated exposure to something *worth* paying attention to.

If you're optimizing only for search, you're building a castle no one visits. You're pouring budget into a channel that may not even be your buyer's first stop anymore. Meanwhile, your competitors are showing up: In the comments. In the summaries. In the short-form videos that AI and algorithms are pushing into feeds.

A blog post optimized for Google might hit page two. But a punchy quote from that same post reshared on LinkedIn with a human voice and a sharp point of view? That gets saved. Commented on. Screenshotted. It lives longer. It lives wider.

This is what it means to build a brand that moves.

It doesn't mean abandoning your website, but it does mean reframing its role. Your website is no longer the entire funnel—it's a validator. A portfolio. A proving ground. People go there to confirm what they've already sensed elsewhere.

So, your job isn't just to publish. It's to *circulate*. To make sure the ideas that live on your website are also:

- Quoted in a newsletter
- Clipped into a 90-second video
- Turned into a LinkedIn carousel
- Referenced in a podcast
- Answered in a comment thread

Distribution is no longer optional. It's the work.

And the companies that figure this out—who publish with a specific voice and repurpose with intent—are the ones being remembered.

The Shift Starts in Your Head

Here's the brutal truth: most companies already have the knowledge to dominate their category. What they lack is the *operating system* to turn that knowledge into consistent content.

That starts with mindset.

Media-driven marketing is not a single department. It's a company-wide posture. It means you:

- Treat content as a strategic asset, not a cost center.
- See internal subject matter experts as talent, not bottlenecks.
- Think in terms of programs, not one-offs.

The tactical shift flows from the mental one and it begins with better questions:

- What's our flagship content format?
- What's our editorial cadence?
- Who owns storytelling inside our org?
- How do we capture and repurpose internal knowledge?
- What would it take to publish like a beat reporter?

The minute you stop asking, "How do we get people to care about our product?" and start asking, "How do we earn trust by being the most useful voice in the room?"—you're in.

You're not chasing relevance. You *are* the relevance.

That's not a campaign mindset. That's a media mindset.

And it's the foundation of everything that follows. If trust starts with showing up, it sharpens with knowing exactly who you're showing up for.

A Tale of Two Teams

Team A: The Campaigners They launch a quarterly campaign. They build a landing page, push it on LinkedIn, and run a few ads. They get a short spike in traffic, a few leads, and then... silence. By the time they regroup for the next campaign, their audience has moved on.

Team B: The Media Operators They publish a weekly LinkedIn series. They send a newsletter every other Tuesday. They record a monthly podcast episode with internal subject matter experts. They reuse, remix, and republish across formats. Their audience knows when to expect them, what they sound like, and why it's worth paying attention.

One team is chasing attention. The other is *earning* it.

Quick Win: How To Make "Boring" Content Actually Work

- Zoom in: Highlight one operator, one project, one breakthrough.

- Make it visual: Shoot a 60-second walkthrough or dashboard breakdown.

- Tie it to a question: "How do I know if my system is optimized?"

- Turn that answer into a recurring series.

Notes

2

Know Your Audience Like a Journalist

Before you publish anything—before you hit record, hit send, or write a single headline—you need to know exactly who you're speaking to and what they care about. And not in a "B2B persona card" kind of way.

We're talking about real, lived-in audience knowledge. The kind journalists use when they're assigned a beat.

Because if you're serious about building a content engine that earns trust, then your job is to know your audience like a beat reporter. Not just their demographics, but their day-to-day. Their pressures. Their politics. Their private Slack messages.

Audience Segmentation vs. Beat Reporting

Marketers love segmentation. They slice and dice audiences into personas, verticals, buying stages, firmographics. That's all fine. But most personas are paint-by-numbers.

They give you surface traits: "Operations Olivia is 38, cares about efficiency, and reads industry news."

Cool. Now what?

Journalists, by contrast, don't stop at the surface. When a reporter covers the education beat, they don't just know that teachers are busy and underpaid. They know which issues are spiking tensions at school board meetings. They know the local union reps. They know the open secrets, the shared frustrations, the things people won't say on the record but always say off it.

That's the depth you need.

Good reporters spend time in the field. They show up at meetings, listen to people complain, and observe patterns. They learn which phrases get repeated, which tensions simmer under the surface, and which questions never get satisfying answers. Then—and only then—do they start writing.

You don't need 10 personas. You need three buyer beats:

- The decision-maker

- The end user

- The skeptical influencer

And for each one, you need to know:

- What they're trying to prove

- What they're trying to avoid

- Where they spend time online

- Who they trust

- What questions they ask in meetings but never on social

Here's the trick: Stop trying to fill out a persona template. Instead, build a "beat file"—a living doc with notes, quotes, examples, and sources for each audience type. Update it weekly. Treat it like your own internal newsroom.

Example: The Engineer vs. The Procurement Officer

Say you're marketing industrial HVAC systems. Your content for an engineer should sound like it was written by someone who's stood on a ladder troubleshooting airflow sensors. It should reference the nuances of system design, retrofitting, and installation timelines.

But your procurement content? Totally different animal. It needs to be focused on risk mitigation, lifecycle cost, and vendor evaluation checklists. It should give them language they can copy and paste into their internal approval docs.

Same product. Two beats. Two very different stories.

This is what journalists call "writing to your source." You're not publishing a general-interest piece. You're writing *for someone*, not just *about something*.

Why Interviewing Your Customers Is a Cheat Code

There is no faster, more efficient, and more underrated way to sharpen your content strategy than talking directly to the people you're trying to reach. Not surveys. Not NPS scores. Not focus groups moderated by someone who last read your blog in 2019. Real, unscripted conversations.

A single customer interview can yield more usable insights than a month's worth of analytics. Why? Because it doesn't just give you data. It gives you *language*. It gives you *context*.

It gives you a window into how your buyers actually think, talk, and make decisions.

When a customer tells you, "The onboarding experience felt like a cold shower," that's not a sentiment score. That's a *headline*. That's copy. That's raw material you can repurpose into value statements, landing pages, blog openers, and product updates. You don't get that from a pie chart.

Great interviews are not about extracting testimonials. They're about pattern recognition:

- What metaphors do they use to describe your product?
- What words do they repeat without realizing it?
- Where do they get stuck when describing their pain point?
- Which parts of your offering do they explain better than your website does?

You want to learn what they say when they're not trying to impress anyone. When they're speaking freely. That's where the insight lives.

This is also where your editorial calendar comes from. If a buyer says, "We had no idea which solution made the most long-term sense," guess what? That's a blog post. That's a podcast episode. That's a newsletter headline: *How to Compare Long-Term Value Between Vendors (When Everyone Sounds the Same)*.

How To Run a B2B Interview Like a Journalist

Here's how to make every interview count:

1. **Prep lightly.** Come in with five to six core questions, but don't follow a script. Let the subject lead.

2. **Record everything.** Use a tool like Otter or Descript. Don't rely on notes—you'll miss the nuance.

3. **Open with disarming questions.** Ask about their background, not their pain. Get them comfortable.

4. **Follow the thread.** If they say, "We almost lost a client because of X," don't move on. That's the story.

5. **Ask dumb questions.** "Can you give me an example?" and "What does that look like day-to-day?" unlock depth.

6. **Close with this:** "If you were me, what content would you create to help someone like you?"

These are journalism moves. They're designed to surface what the audience actually needs, not what we assume they want.

After the Interview: Turn Language Into Strategy

The magic doesn't stop when you hang up. It starts when you analyze what you heard.

After the call:

- Highlight quotes that spark ideas.
- Build a "swipe file" of buyer phrases.
- Flag every question asked that you haven't answered in your content.
- Use their pain points to title content pillars.

- Turn the full transcript into a mini brief for your next piece.

Start building your own internal playbook:

- A doc of buyer quotes
- A doc of objections and how they are phrased
- A doc of questions you need to answer

This becomes your editorial well. Return to it every time you hit a creative wall. It won't just help you write better content—it will make your team *smarter* about the audience.

Add a Listening Layer

Journalists don't rely on a single interview. They build a network of sources and a rhythm of listening.

You should too.

Start by tracking the conversations your audience is already having:

- LinkedIn comments
- Reddit threads
- Industry podcasts
- Niche newsletters
- Webinars you didn't host

Create a doc. Dump quotes, questions, and hot takes. Highlight patterns. Review monthly. Let it shape your roadmap.

Treat it like a newsroom morning meeting: What's everyone talking about? What questions aren't being answered? What stories are missing?

Build an Editorial Calendar Based on Buyer Intent

Most B2B editorial calendars are built around company priorities: product launches, event schedules, internal milestones.

Flip it.

Build your calendar around what your buyers are thinking about *month by month*.

- What are their seasonal cycles?
- When do they start researching?
- What conferences do they attend and why?
- What gets them promoted?
- What keeps them from sleeping in Q4?

Use that to plot out your content rhythm:

- January: annual planning tools
- March: regulatory updates
- June: budget optimization playbooks
- August: mid-year trend reports
- October: board meeting cheat sheets

That's not just smart marketing. That's newsroom thinking. You're anticipating the questions before they're asked.

Writing for Readers, Not Leads

You're not writing for a funnel stage. You're writing for a person. A smart, skeptical, overworked person who doesn't owe you their attention.

That means:

- No generic intros ("In today's fast-paced business world...")
- No SEO-stuffed jargon bombs
- No 800-word blog posts that could've been a tweet

Write like a newsletter editor, not a copywriter.

- Lead with the insight, not the setup.
- Say the quiet part out loud.
- Make it skimmable but meaningful.
- Add friction where it counts—challenge the reader's assumptions.

Ask yourself: What's the one thing this person would actually share with their team?

The best B2B content feels like journalism, with a bias toward action. You ever notice that?

Quick Litmus Test

- If your content makes someone say, "I could've written this," it's not specific enough.
- If it makes someone say, "How did they know this is exactly what I needed?"—you nailed it.

Action Steps

1. **Pick three audience beats.** Don't build 12 personas. Go deep on three buyer types you need to win.

2. **Interview five customers.** Record, transcribe, and highlight their actual words. Find patterns.

3. **Build a listening doc.** Track where your audience talks. Mine it monthly.

4. **Map content to real questions.** Write headlines based on actual audience phrasing.

5. **Review your top 10 pieces.** Are they company-first or audience-first? Rewrite two with the reader in mind.

Knowing your audience isn't a research phase. It's a muscle. One you build by showing up, listening hard, and publishing with empathy.

When you know who you're writing for, everything gets sharper:

- Your voice
- Your headlines
- Your formats
- Your value prop

Most importantly, your reader feels it. And they come back.

That's how you build trust. That's how you build audience. And that's the first job of any media brand.

Knowing your audience isn't enough. You have to turn that insight into content that feels like part of their experience. That process, that feedback loop, is central to building an owned audience, which, as we mentioned at the very start of this book, is where media brands excel.

There's a difference between publishing to the feed and building a loyal readership. One is ephemeral. The other is resilient. And in 2025, resilience is everything. Social platforms change. Search behavior shifts. AI tools scrape and summarize. If you're relying on someone else's algorithm to reach your market, you're always a step behind.

The answer? Build your own pipeline. Convert casual attention into long-term engagement. Not just traffic—but *subscribers*. Not just readers—but *repeaters*.

Stop Thinking in "Views." Start Thinking in "Subscribers."

Marketers are used to chasing numbers: pageviews, likes, shares. But vanity metrics don't build relationship equity. They're sugar highs. What you need is a subscriber-driven approach to your brand, a clear plan for capturing the people who raise their hand and say, "This is for me."

Your editorial planning should include a conversion layer. Not a mere pop-up. A strategy.

That means every blog post, LinkedIn article, podcast, or webinar needs to ask: "What's the next step this reader could take to stay in our orbit?"

It could be:

- A low-friction newsletter opt-in
- A "follow the show" CTA on LinkedIn
- A bonus resource that trades value for email

Smart brands don't just hit publish. They build entry points, then follow up with rhythm, voice, and value that keeps people around.

Build Feedback Loops Into Your Publishing Cadence

Owned audience growth isn't just about accumulating emails --it's about tightening your editorial feedback loop. Everything we discussed earlier in this chapter (the journalistic elements of content marketing) becomes so much easier as you step into this process of building an owned audience through feedback.

The best newsletters, podcasts, and blog series are shaped by the audience. They ask for replies. They highlight reader wins. They cite audience questions in the next issue. This does two things at once: it increases engagement *and* gives you better data.

Here's a simple format that works:

1. **Ask a question at the end of every newsletter.** "What's the one metric you wish your team tracked better? Hit reply."

2. **Log the responses.** Create a running Google Doc or Notion database of questions, terms, anecdotes, and rants. These become your editorial roadmap.

3. **Close the loop.** When you publish something based on audience feedback, tell them. "You asked, we answered." It signals that you're listening—and that this is a conversation, not a broadcast.

This is how owned media becomes earned attention. You stop guessing. You start steering.

Convert "Light Touch" Into Deep Signal

A like on LinkedIn is not the same as a newsletter signup. A YouTube view is not the same as a listener who downloads every episode. You don't need everyone; you need the right people to come closer.

To do that, you need a conversion flywheel that turns public value into private connection.

Example:

- You post a sharp 90-second video on LinkedIn.
- At the end, you say: "Want the full framework? It's in this week's Field Notes. Drop your email and I'll send it."
- The newsletter includes a bonus template and asks for a reply.

That's a loop. You've taken something short, memorable, and public—and, with consent, turned it into a relationship.

Don't optimize for virality. Optimize for depth. Depth compounds.

Insulate Yourself From the Algorithm

This is the real reason to build an owned audience. You're protecting your ability to communicate on your own terms.

Google might stop sending traffic. LinkedIn might throttle reach. AI tools might strip the author's name from your insights.

But if your audience opens your newsletter every Tuesday or listens to your show on their commute, you're insulated.

You're no longer dependent. You're direct.

That's the kind of marketing that doesn't get cut in a downturn. That's the kind of brand that gets invited to the RFP before it's public.

Jim's Thoughts

Interviewing Builds More Than Content

Interviews don't just give you intel. They deepen customer relationships. They signal that you care about their perspective, not just their purchase order. They create advocates, not just users.

We've seen teams book interviews for content and walk away with:

- Customer quotes that become testimonials
- Product feedback that shapes roadmaps
- Warm intros to new prospects
- Executive buy-in on thought leadership initiatives

And when those interviews get turned into published stories—field notes, Q&As, short videos—they become proof. Your brand is seen as a listener, not just a talker.

Notes

3

Content Isn't a Department. It's the Product

In most companies, content lives in the marketing department—wedged between brand decks, campaign calendars, and a dozen competing priorities. It's seen as support. Collateral. An accessory to the "real" work of selling.

That's the wrong frame.

Content *is* the product—at least in the mind of your future customer.

It's the first experience they have with your brand. The first answer they read. The first expert they hear. The first signal that says, "This company gets it."

If the content is thoughtful, helpful, and well-made, the brand earns credibility. If the content is thin, generic, or outdated, you lose trust before you even know their name.

And here's the truth: no one's waiting for your pitch deck. They're judging your value by what you publish, how often you publish, and whether it's actually useful.

Let's reframe how we think about content.

Content Is the Customer Experience Before the Sale

Your product might be outstanding. Your sales team might be elite. But if your content is weak, none of that matters—because most prospects won't stick around long enough to find out.

In B2B, the majority of the buyer's journey happens before they talk to anyone on your team. That journey isn't made up of sales calls. It's made up of micro-moments: search queries, newsletter scrolls, podcast listens, social media fragments, and YouTube recommendations. That's the new front door.

Which means your content *is* the customer experience. Not adjacent to it. Not in support of it. *It is it.*

Let's make this real:

- **Blog posts are product demos**—delivered through story and insight, not slide decks.
- **Webinars are sales calls**—without the pressure or the small talk.
- **Social posts are micro-pitches**—earned attention in exchange for relevance.
- **Newsletters are onboarding flows**—for people who haven't bought yet but are curious.
- **Videos are trial runs**—they let prospects test-drive your expertise in 60 seconds.

Every piece of content you publish is a signal: You can trust us. We understand your world. We solve the kinds of problems you're facing.

And here's the kicker: Every piece that falls flat sends the opposite message.

Content as First Contact

When someone first hears your company name, they're not booking a call. They're Googling you. They're checking if you've written about their problem. They're skimming your LinkedIn. They're seeing if your voice feels like someone who *gets it*—or just another vendor selling digital wallpaper.

If your content is vague, outdated, or hard to navigate, they won't tell you. They'll just leave. And the opportunity is gone before your sales team ever got a chance.

That's why your content isn't a warm-up act. It's the *main event* for 90% of your audience.

Think Like UX Designers

User experience professionals obsess over "first touch friction." They ask, "What's the experience of someone landing on this page for the first time? What do they see? What's the first interaction they'll have? Does it build trust or create doubt?"

You need to bring that same lens to your content.

Look at your content footprint the way a buyer would:

- What's the first blog post they'll find?
- What does your CEO's last LinkedIn post say?
- What video shows up in a YouTube search?

- What does your email newsletter look like to someone who just subscribed?

If any of that content is confusing, irrelevant, or self-centered, that's a broken experience. That's a prospect bouncing off your brand—before they even realize what you do well.

The Silent Funnel

Here's the part most companies underestimate: A huge number of deals are lost in *silence*.

Not because you didn't follow up, but because the prospect made their decision before you even knew they were interested.

They visited your site. They watched a video. They read one blog post that could've changed their mind—and it didn't. Because it was too fluffy. Too product-heavy. Too generic. Or just published in 2021 and never updated.

And they moved on.

They never told you. You never knew.

That's the silent funnel. And your content is either building momentum inside it or quietly repelling people from it.

Great Content *Feels* Like Working With You

Here's the standard you're aiming for: *Your content should feel like a preview of what it's like to work with your company.*

It should demonstrate how you think. How you solve problems. How you talk to customers. It should reflect your standards—your clarity, your specificity, your style of leadership.

A great piece of content isn't just informative. It's experiential.

That's why it matters that your writing is clear. That your videos are well-lit. That your webinars are run with purpose. Because these things aren't just marketing collateral—they're *simulations of partnership.*

Would you trust a vendor who can't explain their product clearly in an article?

Would you feel excited to work with someone whose podcast rambles and never lands a point?

Would you book a demo after watching a demo that feels like a snooze button?

Exactly.

Build for the Buyer's Mindset, Not the Marketing Funnel

Most buyers don't follow your funnel. They follow their *gut.*

And that gut feeling is shaped by content.

- A sharp LinkedIn post tells them: *These people understand what I'm up against.*
- A case study tells them: *They've solved this before.*
- A webinar tells them: *They have depth. They're not winging it. They know their stuff.*
- A weekly newsletter tells them: *They're consistent. Reliable. Present.*

That's not marketing fluff. That's how trust forms.

Every Touchpoint Is a Trust Test

Here's the core idea: Every single touchpoint is a chance to build trust—or break it.

Most companies don't lose deals because their product is bad. They lose them because their *content doesn't earn the next click.*

It doesn't show up when the buyer is looking.

It doesn't say what the buyer needs to hear.

It doesn't sound like someone worth listening to.

So the buyer drifts. Not because they said no—because they never said anything at all.

Content *Is* the Experience

We need to stop treating content like decoration. It's not a garnish. It's the entrée. It's the thing that creates gravity around your brand.

The blog is the demo.

The video is the sales call.

The newsletter is the relationship.

Every single piece is a moment of truth. A moment to show your audience that you're not just a vendor—you're the one who actually *gets it.*

When you do that, you don't have to convince people to enter the funnel. They're already halfway through it—because your content brought them there.

Example: HubSpot's Content Flywheel

HubSpot didn't grow because of a giant ad budget.

They grew by publishing an *insane* amount of content that ranked, answered real questions, and positioned them as the authority on inbound marketing.

By the time someone was ready to evaluate tools, they'd already spent hours learning from HubSpot. The trust was already banked.

Your brand should be building the same kind of pre-sale compound interest.

Marketing and Sales Alignment Through Editorial Thinking

"Marketing and sales need to be aligned." Great. How?

Try this: think like an editorial board.

In a newsroom, reporters don't work in silos. The editorial team plans coverage around beats, news value, and reader interest. Everyone's rowing in the same direction.

Your sales team is your reporting team. They hear the objections, the language, the misconceptions. Your marketing team is editorial—they shape that raw insight into useful stories, guides, and resources.

But it only works if they're in the room together.

Practical Tactics:

- **Hold monthly editorial syncs.** Bring sales and marketing together to surface what prospects are asking. Build content from those questions.

- **Create a shared content backlog.** Let sales reps drop in common objections or content gaps. Marketing turns them into assets.

- **Use enablement assets externally.** A sales battle card can become a blog post. A customer onboarding doc can become a how-to article.

Build Your Content Portfolio Like a Newsroom

Most B2B content calendars are spreadsheets. They list formats, deadlines, and CTAs. That's fine for logistics, but it's not how great content gets made.

Newsrooms operate differently. They think in beats, formats, and anchors.

Here's how to borrow their model:

1. **Define your beats.** These are the core problems and topics your audience cares about. Each beat gets an owner and a strategy.

2. **Create recurring formats.** Newsletters, Q&A videos, behind-the-scenes series—formats that are recognizable and repeatable.

3. **Establish anchor pieces.** Long-form, evergreen content that acts as a hub for related pieces.

4. **Layer with fast-twitch content.** Timely takes, LinkedIn posts, newsletter riffs—these create relevance and momentum.

This layered system turns your content from a flat calendar into a living, breathing media engine.

Example: Duolingo's Content Stack

On TikTok, Duolingo runs a recurring video series with their owl mascot. On YouTube, they publish longer-form language tutorials. Their app content is constantly refreshed. And their social team engages in real-time trends. It's not chaos—it's layered content by intent and cadence.

B2B can do the same. It just takes structure.

Internal Knowledge as Proprietary Media IP

Your company is sitting on a goldmine of untapped media IP. It lives in sales calls, Slack threads, client emails, onboarding docs, and the brains of your subject matter experts (subject matter experts).

Stop treating that knowledge like tribal lore. Start turning it into assets.

How to Operationalize Subject Matter Expert Content:

- Interview subject matter experts on Zoom. Chop up the audio/video for posts.
- Ghostwrite articles under their byline. Use their real voice.
- Create "notes from the field" content based on real client stories.
- Host internal show-and-tells. Turn those sessions into public-facing content.

Final Takeaway: Build Trust Like a Product

If you treat content as a support function, it will underperform. If you treat it like the product your audience experiences first, everything changes.

- You prioritize quality.
- You develop systems.
- You seek feedback.
- You measure value by impact, not just clicks.

Most importantly, you stop waiting for permission to lead the conversation. You start acting like the publication of record for your niche.

In B2B, the brand that earns trust early wins long.

Content isn't a department. It's your storefront, your voice, your reputation.

Treat it like the product it is.

And yet, content without distribution is like a signal without a receiver. It's time to build your delivery system.

―――――――――― C ――――――――――

Scott's Thoughts

Your Sales Team Is the Field Reporter

We talk a lot about "marketing and sales alignment." But most advice stops at vibes.

Here's a sharper lens: Treat your sales team like field reporters.

In journalism, the best stories start in the field—not the newsroom. Reporters bring raw observations, quotes, and emerging themes back to editors. Editors shape those into coverage that informs and resonates.

Apply that model to your business:

- **Sales = reporters.** They're on the front lines. Hearing objections. Watching body language. Noticing where prospects pause, get confused, or lean in.

- **Marketing = editorial.** They turn that field intel into high-quality stories, explainers, guides, and resources.

But here's the catch: most companies treat these teams like separate departments, so the intel never makes it back. The questions that sales hear never get answered in content. The objections never get preempted. The words customers use never shape the messaging.

Want to fix that?

- Have your top AE drop in on your next content brainstorm.
- Review lost deals and ask, "What content could've saved this?"
- Turn every sales call into a quote library. Real words. Real people. Real leverage.

When sales informs editorial, your content stops being theoretical—and starts feeling *uncannily useful*.

That's alignment. Not just because you say so—but because you *publish like a team*.

Notes

4

Build Your Distribution Muscle

L et's get one thing straight: If your content isn't being seen, it's not working. Period.

This is the hard truth most marketing teams eventually face. They spend weeks perfecting a blog post, webinar, or video—only to see it fade into obscurity. Not because it was bad content, but because they didn't know how to distribute it.

Distribution isn't the last step. It's half the job.

If content is the product, distribution is your delivery system. And if you want people to actually *consume* what you're creating, you need to build serious muscle here.

Why "If You Build It, They Won't Come"

The biggest myth in content marketing is that great content markets itself. It doesn't. Not in a world where attention is fragmented, audiences are algorithm-gated, and people are drowning in noise.

A beautiful piece of content sitting on your blog without distribution is like a billboard in the desert. Technically there, but not seen. Not engaged with. Not remembered.

You can't afford to publish and pray. You have to *publish and push*.

This means rethinking how you:

- Plan content
- Package content
- Share content
- Re-share content
- Route content internally and externally

It's not about working harder. It's about working smarter, repeatedly.

Build Internal Distribution Workflows

If you want to build a distribution engine that runs consistently, you have to start internally. Most content fizzles out not because it lacks quality but because no one takes ownership of getting it into the world.

Your solution? Treat distribution like a team sport with defined plays and clear ownership.

Step 1: Create a Content-to-Distribution Pipeline

This is your internal relay system. It ensures every asset created flows through a standardized set of remix, review, and redistribution steps.

Example Workflow:

1. **Content is finalized.**
2. **Repurposing plan is drafted.** (Three to five short-form snippets, one to two visual assets, social captions)
3. **Distribution brief is created.**
4. **Assets dropped into Slack or Notion for review.**
5. **Team shares through assigned channels.**
6. **Posts scheduled, reshared, logged.**

Step 2: Assign Roles With Names, Not Departments

- **Distribution Lead:** Owns the final step of pushing live and scheduling reposts.
- **Subject Matter Expert Liaisons:** Coordinates with execs or reps to personalize and share content.
- **Content Strategist:** Writes modular content with distribution in mind.
- **Analytics Lead:** Reports on what worked and what didn't.

Don't let distribution die in a generic email thread. Make it visible. Make it shared. Make it repeatable.

Step 3: Build Distribution Into Project Planning

Every content brief should answer:

- What channels will this show up on?
- How will it be sliced and repurposed?
- Who needs to be involved in distribution?

- What does success look like?

By asking these upfront, you force teams to consider distribution *before* they ever write the first word.

Calendarized Distribution: How To Extend Content for 30 Days

You don't need more content. You need to use what you've already created in smarter, more sustainable ways.

Here's a sample 30-day distribution plan for a single blog post:

Anchor Asset: Blog Post on "How Greenhouses Can Cut Energy Costs in Q4"

Week 1

- Publish blog post and share on company LinkedIn
- Record a 90-second video summary with subject matter expert and post to YouTube and LinkedIn
- Email the blog post to your newsletter audience with added commentary

Week 2

- Turn three blog insights into standalone quote cards
- Have the CEO post one takeaway with a personal POV
- Repost the original blog on LinkedIn with a new hook

Week 3

- Cut the subject matter expert video into three clips and post to Instagram Reels + X
- Send a follow-up email offering a downloadable checklist based on the blog
- Publish a poll on LinkedIn asking which tactic readers use most

Week 4

- Turn the blog into a SlideShare for LinkedIn
- Share a behind-the-scenes image of the subject matter expert video recording
- Publish a Q&A-style follow-up post: "You Asked, We Answered"

Each move extends the shelf life of the original idea and gives your audience multiple entry points.

Pro Tip: Use a single spreadsheet or project tracker to manage this cycle. Every piece of content should have a 30-day plan. That's how media brands build visibility *without burning out.*

Use a Distribution Brief for Every Asset

To make this process frictionless, build a five question distribution brief into your content workflow.

The Distribution Brief

1. **What is the core message?**
    - The sticky takeaway you want repeated

2. **Where will this live first?**

   ◦ Blog? YouTube? Podcast? Social?

3. **How will it be sliced and reused?**

   ◦ Break into quote cards, short videos, polls, repromoted email segments

4. **Who is responsible for each leg of distribution?**

   ◦ Assign names, not teams

5. **How will we measure resonance?**

   ◦ Think beyond clicks: saves, comments, sales mentions, repeat engagement

By filling out this simple framework *before* publishing, you'll make your content work 10x harder without adding 10x the labor.

Distribution isn't a postscript. It's the multiplier.

Think in Channels: Media, Not Just Platforms

Most marketers think in platforms: LinkedIn, email, YouTube, podcasts. But media companies think in *channels*. Channels are strategic. They have an audience, a cadence, a format, and a purpose.

Here's a breakdown:

Platform	Channel (Media Mindset)
LinkedIn	Weekly founder POV series
Email	Biweekly insight-led newsletter
YouTube	Monthly Q&A show with internal subject matter experts
Podcast	Industry-specific interview series

Notice the difference? One is a tool. The other is a *show*.

The goal is to treat your distribution touchpoints like programs—things your audience looks forward to, not one-off drops.

Example: CEO Diaries on LinkedIn

A B2B SaaS company launched a weekly LinkedIn post from their CEO called "Five Things I Learned This Week." It wasn't flashy, but it was honest, smart, and consistent. Engagement climbed. Followers doubled. Sales reps started using it as a conversation starter. That's the power of turning a platform into a recurring channel.

Rented Land vs. Owned Land—and How To Use Both

"Don't build your house on rented land." Sound advice. But incomplete.

Yes, you want to own your audience, but rented channels like LinkedIn, YouTube, and Instagram are where *new* people discover you. They're the front door. Owned channels—like your website, newsletter, and podcast feed—are where they *stay*.

Here's how to balance the two:

- Use rented land to capture attention.
- Use owned land to deepen the relationship.
- Build bridges from one to the other (email CTAs, retargeting, exclusive offers).

This isn't either/or. It's a relay race. Each channel hands off to the next.

Actionable Framework:

1. Create an anchor asset (e.g., deep-dive blog post)
2. Break into three to five snackable formats (Tweets, quote cards, short videos)
3. Distribute across rented platforms
4. Drive to a subscribe/conversion opportunity on owned land
5. Measure by resonance and repurpose again

Leverage Internal Subject Matter Experts and Execs as Distribution Channels

Your internal experts are underused media assets.

Most companies rely solely on the marketing account to distribute content. That's a mistake. Your salespeople, engineers, customer success leads, and execs have networks, credibility, and unique voices. Use them.

Tactics:

- Create "done-for-you" social post templates for key team members

- Film internal conversations and turn them into content snippets
- Host a monthly "expert take" post written or recorded by a different team member
- Let subject matter experts run AMA (ask-me-any-thing) sessions on social

It's not about forcing your team to become influencers—it's about enabling them to contribute in ways that feel authentic and sustainable.

Create a Culture of Re-Sharing

Most teams treat content like a single-use asset: They share it once and move on.

Big mistake.

Media companies repackage, repost, and redistribute constantly. Because they know: just because *you* shared it once doesn't mean your audience saw it.

Build a system to:

- Repost evergreen content on a 90-day rotation
- Update and re-promote old blog posts with new context
- Share popular posts from your team's personal profiles
- Create content series from existing high-performing assets

Pro Tip: Set up a "content remix tracker" where you log every asset and all the formats it can be spun into (e.g., post, clip, quote, visual, story). Assign owners and republish dates.

Build Distribution Into the Content Process

Distribution isn't something you tack on at the end-- it has to be baked into the content creation process itself.

Ask at the start:

- Where will this live?

- How will it be sliced?

- Who will share it?

- What does success look like?

This forces you to:

- Design modular content (e.g., blog + video + postable quotes)

- Write with headlines that work on multiple platforms

- Create with repurposing in mind

If you're making a podcast, record a behind-the-scenes clip. If you're writing a long blog post, highlight key stats for social. Always think: *How can this live in five places, not just one?*

Action Steps

1. **Audit your current channels.** Where are you showing up? Where are you missing?

2. **Design one flagship distribution channel.** Weekly newsletter? LinkedIn POV series? Choose one and build it like a show.

3. **Assign distribution roles.** Who owns reposting? Who monitors subject matter expert participation? Treat it like ops.

4. **Build a remix system.** Start logging every asset and how it can be reused.

5. **Measure differently.** Don't just look at clicks. Look at comments, saves, mentions in sales calls, etc.

Distribution isn't about shouting louder. It's about showing up smarter. More consistently. More intentionally.

The content graveyard is filled with great ideas that no one ever saw.

Your job is to build the system that ensures your content *finds its people.*

And if you do it right, you don't just get more eyeballs.

You get more trust. More traction. More leverage.

Distribution matters a lot, but so does the rhythm of your work. Sporadic content doesn't build brands, series do.

Notes

5

Think in Series, Not One-Offs

One of the biggest traps in content marketing is the "one-and-done" mindset. This rut comes about from those endless marketing meetings where team members brainstorm siloed topics for the next piece of content; no coherence ever emerges because the plan is approached anew in each meeting. You write a blog. You host a webinar. You publish a video. Then you move on to the next thing.

Rinse, repeat, burnout.

This scattershot approach not only exhausts your team-- it also weakens your brand's presence. It forces you to constantly reinvent the wheel instead of building momentum.

Want to fix that?

Think in series.

Why Content Fatigue Is a Sign of Weak Systems

If your team is constantly overwhelmed by content demands, it's not because you're doing too much. It's because you're doing it inefficiently (and likely in isolation, which in turn generates more inefficiencies).

Most content fatigue comes from:

- Treating every piece of content like a new invention
- Lacking reusable formats or templates
- Operating in campaign mode, not programming mode
- Always reacting, never planning

Thinking in series solves all of this. It reduces creative friction. It increases speed. It compounds audience trust. But first, let's zoom out and understand *how* we got here in the first place. Surely this isn't the first time you've heard about burnout in the context of a small content marketing team.

Since 2020, the way we work has changed dramatically. Distributed teams are now the norm. Slack has replaced the newsroom. Editorial calendars are maintained in Notion boards that no one checks. And most content professionals—writers, designers, producers—are operating in fragmented silos, juggling disconnected campaigns with too few inputs and too many expectations.

Burnout is often labeled as a resourcing problem, but it's more of a systems problem. And, more specifically: a collaboration problem.

When content strategy lives on one person's shoulders, every new asset becomes a mini-crisis.

You brainstorm alone. You guess what sales needs. You chase subject matter experts for input. You rewrite to make it sound on-brand. You publish. And then you do it all again next week.

It's no wonder teams are exhausted. This isn't sustainable publishing. It's reaction masquerading as strategy.

The post-pandemic shift toward digital-first work—while flexible—has made this worse in many orgs. Remote teams are struggling to maintain shared context. Meetings are replaced by Slack pings. And content loses its rhythm because nobody feels the rhythm. The result? Stress, silence, or scattered output.

Does this sound familiar? (Rip this page out of the book and mail it to your manager.)

Let's be clear: From the jump, a simple shift in mindset solves a lot of these problems. Thinking in series is the shift you need.

Instead of constantly asking, "What should we publish next?" you get to ask, "What's our next episode?"

Creating Repeatable Content Formats

Media companies don't start from scratch every day. They develop shows, segments, columns, and beats—repeating structures that audiences come to recognize and trust.

You should do the same.

Step 1: Choose a Format Type

- **Q&A Series:** Internal experts answer real customer questions
- **Field Notes:** Lessons learned from recent projects or client work
- **Founder Logs:** Weekly reflections from leadership
- **Explainer Series:** Break down one core topic over multiple episodes
- **Behind-the-Scenes:** A recurring look at how your team actually works

Step 2: Give It a Name

Naming your series gives it an identity. It signals to your audience that this is a repeatable experience.

Examples:

- "Ask the Engineer"
- "Notes from the Field"
- "CEO Diaries"
- "The Briefing: What You Missed This Week"

Make it sticky. Make it sound like something they can subscribe to.

Step 3: Build the Template

- Same format each time (e.g., three questions, 500 words, same intro)
- Branded visual asset or layout
- Set publishing cadence (weekly, biweekly, monthly)

Templates reduce lift. Consistency builds rhythm. Rhythm builds trust.

Editorial Ops: Who Owns What in a Content Series

A series lives or dies on internal ownership. If everyone owns it, no one owns it. Treat each series like a product with a clear team. We'll get into how to develop an internal media engine in Chapter 7, but for now it's important to reiterate the importance of collaboration. However you build your team, with whatever boot-strapped resources you can gather, collaboration is an essential ingredient. Otherwise, you're just back to burnout in three months.

Burnout thrives in isolation. Creative energy fades when every asset feels like a solo act.

But when you build your content strategy around repeatable series—with clear roles, a shared calendar, and a reliable flow of input across departments—you begin to create content with real momentum. The work actually becomes easier and, dare we say, more fun for everyone.

A strong content series becomes a container for collaboration. It gives your sales team a place to contribute ideas. It gives your subject matter experts a predictable way to show up. It gives your designers a style to work within. It gives your writers a cadence they can trust.

And most importantly: it keeps the entire marketing team from spiraling into reaction mode.

Let us suggest an ownership matrix:

- **Series Producer:** Oversees planning, production, and cadence

- **Host/Subject Matter Expert:** On-camera or on-mic contributor
- **Content Lead:** Edits, formats, packages episodes
- **Distributor:** Publishes on core and support channels
- **Analyst:** Measures performance, reports monthly

Then, build an editorial rhythm:

- Monthly sync to plan upcoming episodes
- Quarterly "postmortem" to review what landed and what didn't
- Create a shared doc of content ideas, audience questions, and future episode hooks

A good series doesn't happen by accident. It's maintained like an engine.

How To Know if Your Series Is Working

Not every series hits right away. But if it's hitting at all, you'll feel it in these signals:

Early Indicators:

- People *comment* or *reply* to say, "This is exactly what I needed."
- Colleagues share it internally without being asked.
- It becomes a talking point in sales or customer calls.

Trackable Metrics:

- **Completion rate** (video/podcast)
- **Reply/comment rate** (newsletter/social)

- **Repeat engagement** (people returning episode to episode)
- **Sales mentions** (get sales to log references in deals)

Strategic Metric:

- **Reuse velocity:** Are other teams pulling from this content to use elsewhere? (Sales decks, onboarding, customer FAQs, etc.) The first time a sales rep asks for the link to a recent podcast episode, you'll know you've done it right.

A great series earns attention, yes, but the KPI is in earning reuse.

This is different, and maybe you already feel a bit of internal pushback coming. (That's not an easily trackable metric!) Publishing a series (and publishing effective content marketing pieces in general) is about relationship-building. It's about trust. You don't track the trust you've built with your old college friends, do you? But you know where you stand, you know what you've built over the years and what it all means for your life.

All right, let's zoom back in to content marketing. When you commit to a recurring format—weekly LinkedIn drops, monthly Q&A emails, a three minute Friday podcast—you're now opening a channel. And the smartest brands don't just broadcast through that channel; they listen through it. They adapt. They engage. It's a two-way channel!

This is the difference between content as output and content as conversation. Consider the parasocial relationships that emerge from the best podcasts; listeners feel like the host is their friend. The same goes for a good LinkedIn show: Your audience will come to relate to the host in a way that builds trust and, eventually, maybe, ushers in a sale or some other measure of brand loyalty.

So, if you want your series to actually land—and grow—don't just measure performance from the outside. Cultivate connection from the inside.

Here's how:

1. Signal-Scan Like a Producer

Not every series hits right away. That's normal. But if it's hitting at *all*, you'll feel it—in the margins.

Watch for early engagement signals:

- Someone comments "This is exactly what I needed" or "This tracks with what I'm seeing in the market."
- A sales rep forwards it to a prospect—without you asking. Time to pop some champagne!
- A buyer references it in a Zoom call: "I read something you posted the other day..." You're going to feel like a hero when this happens.
- Internal teams drop it into Slack: "Use this in your next deck."

These are your creative radar pings. They tell you which topics are resonating, which tone is landing, which formats feel natural for your audience. Those radar pings aren't easy to *track*, exactly, but they're the important metric you should orient some measure of success around.

Track them to the best of your ability (even a few notes in your phone will impress your manager at the end of the month). Quote them. Use them to evolve the series in real-time.

2. Ask for Interaction Clearly and Consistently

You don't need a CTA in all caps. You just need an invitation to respond.

Smart ways to prompt audience feedback:

- "Which of these is most useful in your world right now?"
- "Have you seen this challenge in your own org?"
- "What should we unpack in the next issue?"
- "Reply with your take—we'll feature a few in the next episode."

Bonus: Use a recurring engagement device—something your audience expects and can plug into: a monthly "Reader Shoutout," a featured audience quote in your newsletter, a quick poll or reply prompt ("This or that?").

When people see their names, their questions, or their phrasing show up in the series, it builds emotional ownership. And that's how content becomes community. You'll notice, too, that this becomes much easier the further you go along in the process. This is a content-driven authority flywheel (more on that in Chapter 9).

3. Use DMs and Replies as Research, Not Chatter

Most marketers ignore DMs and replies unless they're leads. Big mistake. Smart media thinkers treat every message like intel.

If someone replies to your newsletter, don't just say "Thanks!"

Ask: "What stood out to you?" "Was that a current challenge or one you've already solved?" "Would it be helpful if we turned this into a short guide?"

Every reply becomes a creative breadcrumb. Every DM is a mini-interview. And, in the process, you're collecting:

- Real language for future posts
- New topic ideas from real-world pain points
- Credible quotes for case studies or thought leadership

To reiterate the message from the start of this chapter: the best content calendars aren't made in planning meetings. You'll burn out if you rely on those spaces for your actual output plans. Instead, calendars are built-in inboxes. The feedback loop is your friend.

4. Build Feedback Loops Into the Format Itself

Your series doesn't just have to deliver content. It can react to content. Some ideas:

- Turn reply threads into a new episode ("Here's what our readers said about last week's piece...")
- Run a "community question" segment every third post
- Dedicate one newsletter per quarter to top-performing topics—*and the ideas your audience wanted more of*

If your series starts to talk back to your audience, you've hit a different level. Now you've got a trusted platform, in much the same way that your local newspaper or your favorite YouTube channel established trust.

5. Measure More Than Metrics

Reach is easy to track. But real engagement shows up in:

- Bookmarked posts
- Screenshots shared in Slack threads
- Comments like: "Forwarded this to my whole team"
- Repeat opens or listens from the same domain

This is going to be a hard sell to your team, but we recommend setting up a "memory swipe file" internally. Every time your content is quoted, referenced, or re-used by your audience, drop it in.

Over time, you'll build a living case study of what actually connects.

–––––––––––––––––– ◖ ––––––––––––––––––

Series Shelf Life: When To Kill, Rerun, or Refresh

Not every series lives forever—and that's okay. Here's how to know what to do next:

Retire the Series When:

- You've seen upwards of 10 episodes underperform after a refresh
- The original problem or audience has shifted

- The internal owner no longer has bandwidth or buy-in

Evolve the Series When:

- Engagement has plateaued but core interest remains
- A new medium could breathe life into it (e.g., blog → podcast)
- Feedback points to a stronger format or tone

Rerun the Series When:

- Content is seasonal or event-driven (e.g., "Budget Month," "Year-End Wraps")
- You have new audience segments unfamiliar with the original posts
- A repromoted run helps feed your current distribution cycle

Every content program needs a content lifecycle. Treat your series like a living asset, not a forever promise.

How Episodic Content Builds Audience Memory

Here's the magic of series-based content: it creates mental real estate. Audiences start to anticipate your content. They recognize it in their feed. They remember it. They talk about it. It becomes part of their routine.

Compare these two approaches:

- **One-Off Blog:** "How to Improve Internal Communication in Hybrid Teams"
- **Series Format:** "Hybrid Habits"—a weekly three minute tip for better remote work

Same topic. Totally different energy. The second approach builds brand and value over time.

The goal isn't just to publish. It's to create signature content that *sticks*.

Internal Benefits of Serial Thinking

This isn't just about your audience; it's about your team. When you build a few strong content series:

- Your team knows what to make next
- Your subject matter experts know what's expected of them
- Your audience development team knows what to promote
- Your sales team knows what to reference

It creates internal alignment. It turns content from a task into a system.

Real-World Example: Encore360's Recurring Formats

Encore360 worked with a landscape fabric company to launch a video series called "Tinkering with Bart."

Every week, Encore's content team connected with the client's subject matter expert, who walked us through what was happening in his own garden (using the company's products). Same simple, lo-fi format. Same cadence. New topics each time.

It became:

- A customer engagement tool
- A top-of-funnel lead generator
- A fun opportunity to connect with prospects

All from a single recurring show.

Build a Slate, Not a Stack

You don't need more content. You need better structure.

We're going to reiterate this idea even more in coming chapters, but think like a media planner:

- One flagship series for authority (e.g., expert interviews)
- One recurring internal POV series (e.g., founder notes)
- One tactical series for immediate value (e.g., weekly how-to post)

That's your content slate. Three programs. Three formats. High impact, low chaos.

Everything else flows from there.

Action Steps

1. **Audit your recent content.** What themes are repeating? Where is the spark?
2. **Group content into categories.** Could any become a recurring series?

3. **Pick one idea to turn into a show.** Start small—three episodes, same format.

4. **Give it a name and cadence.** Commit to publishing it at least three to five times.

5. **Build a template.** Standardize visuals, structure, and workflow.

6. **Promote it as a program.** Announce it. Tease it. Get people to look forward to it.

Thinking in series isn't just a publishing strategy. It's a mindset shift.

You stop playing content whack-a-mole. You start building rhythm. You stop chasing attention. You start earning loyalty.

And, over time, your brand stops sounding like a company that wants to be heard and starts sounding like a company your audience can't wait to hear from. Leaving lasting memories for your audience depends on visibility. Let's talk about creating content that people actually remember.

Eric's Thoughts

Why Stickiness Matters More Than Ever

Today's audience isn't reading your content in sequence. They're encountering it in fragments: one quote on LinkedIn, one clip in a Slack channel, one chart screenshot in a sales deck.

Meaning: If your content doesn't stick, it doesn't exist.

That's the power of a content series. It gives your audience something recognizable to hold onto. A name. A format. A rhythm. In a world where attention resets every 30 seconds, that cohesion builds memory.

One-off blog topics feel like noise. Series create signal. One-off content gets consumed. Series get anticipated. One-offs compete for space. Series *claim* space.

The job isn't just to publish something good. It's to be *remembered*.

And memory is built through repetition, structure, and tone.

6

Create for Impressions, Not Just Clicks

Clicks are overrated.

Let's say that louder for the marketers in the back: Clicks are not the metric that matters most.

Sure, they're easy to measure. They make dashboards look busy. But in a media-driven marketing model, what really matters is impressions—and more specifically, impression with impact.

Because before you can get someone to click, subscribe, or buy, you have to get them to stop. To notice. To think. To feel something. And that begins with reach, resonance, and recall, not raw traffic. This fixation on traffic as a KPI is an important habit to break.

We talked about pushing content outward in Chapter 4. Here, we're zooming in on how to package ideas for native impact, even if no one clicks through to your website.

What Media Teams Understand About Attention

Media professionals know something most marketers forget: the majority of content consumption is passive.

- People scroll past headlines.
- They skim posts without clicking.
- They read a newsletter preview but never open it.
- They absorb your content *ambiently*—through repetition, exposure, and memory.

And guess what? That still builds brand.

When done right, impression-rich content:

- Teaches before the click
- Delivers insight at a glance
- Reinforces your voice and POV repeatedly
- Primes trust before action is needed

Why Zero-Click Content Is Good Strategy

Zero-click content is content that delivers value *without* requiring the user to leave the platform.

In traditional marketing, this sounds like failure: "But they didn't convert!"

In media marketing, it sounds like trust-building: "They learned something. They remembered us. They'll be back."

Examples:

- A full answer in a LinkedIn post
- A 90-second video explaining a complex concept

- A visual quote card summarizing a full blog
- A Twitter/X thread that teaches instead of teases

Here's what happens:

1. You earn attention by being generous
2. You create mental availability
3. You become a known entity before they ever click

And when they *do* need what you offer, they don't Google the category. They search for *you*.

Zero-Click Format Library

Zero-click content isn't an aesthetic choice. It's a strategic one. It's how you teach, earn trust, and stay visible *without asking your audience to do anything*. And, when done right, it builds more brand equity than most gated white papers ever could.

Here's a working library of zero-click formats that every B2B brand should have in rotation:

1. The Single-Slide Visual Explainer

Format: One image, one core idea

Use: LinkedIn, Instagram, email, decks

Example: A diagram showing how your product fits into a facility's workflow—*no link needed*. Make it scannable. Branded. Shareable.

Why it works: Visuals don't just summarize—they *travel*. Great diagrams get screenshot, Slack-shared, and remembered.

2. The X/LinkedIn Teaching Thread

Format: 5–10 posts that unpack a core insight

Use: Twitter/X, LinkedIn carousels

Example: "Seven things your procurement team actually wants to know before a system install." No links. Just value.

Why it works: The reader stays on the platform. The insight builds trust. The voice builds familiarity.

3. The Executive POV Post

Format: First-person mini-essay (150–300 words)

Use: LinkedIn or newsletter intro block

Example: "One thing I learned this week running sales for a water tech company..." Direct. Insightful. *No CTA needed.*

Why it works: Builds leadership visibility. Adds humanity to the brand. People buy from people they believe in.

4. The Contextual Quote Card

Format: Branded visual with an original insight

Use: Email, social, internal decks

Example: A quote from your subject matter expert layered on a graph or process screenshot. "Humidity control isn't just a climate issue—it's a sugar production issue." — Lead Grower

Why it works: Quotes are fast to read, fast to share, and packed with subtext. A good quote card says: *We know what we're talking about—and we're not afraid to say it clearly.*

5. The Micro-FAQ

Format: One question, one answer, two to three paragraphs max

Use: Newsletter sections, blog sidebars, video clips

Example: "Why does UV degradation matter in high-altitude installs?" Answer it. Don't tease it. No click required.

Why it works: Anticipates buyer questions and builds "they get me" energy. Great also for sales enablement and internal reuse.

Start building a swipe file of these formats. Keep them short. Keep them punchy. Keep them visible.

And remember: value isn't something you bait people toward. It's something you *deliver*—right where they are.

Common Mistakes in the Shift to Impressions-First

Moving from click-obsessed to impression-smart isn't always smooth. Here are the most common faceplants—and how to avoid them.

Mistake #1: Posting Valuable Content Without Framing It

The idea's good—but the packaging flops. Dense copy, no white space, no headline, no takeaway.

Fix:

Lead with the *insight*, not the context.

Add a bold hook. Use bullets, bolding, white space. Make the *skim* feel satisfying.

Mistake #2: Going Zero-Click but Forgetting To Deliver

You post a "mini essay"...but it's vague. There's no clear point. The reader walks away with nothing.

Fix:

Every zero-click post should answer one question: *What is the reader walking away with?* Teach *something*. Even in two sentences.

Mistake #3: Forgetting to Brand It

You're publishing good content—but it's unbranded, unstyled, and anonymous. The audience may remember the idea, but they won't associate it with *you*.

Fix:

Use consistent visuals. Recurring formats. Branded voice. Even your sentence rhythm and tone should feel like part of a pattern.

Mistake #4: Quitting Too Soon

You try zero-click for two weeks, don't see conversions, and call it a failure.

Fix:

This is a *compound interest* game. You're building memory. Keep showing up. Repeat your POV until your audience can quote it.

Make Your Ideas Portable

Content doesn't live in a vacuum. Your best ideas should travel.

That means every key idea should be packaged in multiple formats:

- A single stat becomes a chart, a Tweet, and a pull quote
- A blog post becomes a video explainer and a newsletter snippet
- A long webinar becomes short clips, takeaways, and quote cards

Think of your content like a song: the chorus should be catchy, repeatable, and singable across formats.

Real-World Example: The Pillar Post System

Encore360 worked with a greenhouse technology company to turn their long-form blogs into what we called "pillar post kits."

Each blog was repurposed into:

- Three LinkedIn posts (quotes + charts)
- Two newsletter blurbs (in some cases picked up by B2B trade pubs)
- One short video summary
- A single slide for sales enablement

The original article was great, but the kit turned it into a distributed media presence. It showed up everywhere.

The Deeper Logic of Mental Availability

Let's talk about *why* impressions-first thinking works—and why the brands that win don't just get remembered...they get *recalled*.

The term you want to steal from brand science is this: Mental Availability.

Mental availability is the probability that someone thinks of your brand in a buying moment. It's not about love. It's about *recall under pressure*.

You know what helps?

- **Repetition**
- **Distinctiveness**
- **Clarity**
- **Familiarity**

This is why someone types "Gong" instead of "revenue intelligence platform."

It's why they search "Basecamp" instead of "project management software."

It's why they ask for "Tinkering with Bart" instead of "Can you send me that gardening video?"

That's the win.

Your job is to become the *default* for your category. That means:

- Showing up *before* they need you
- Speaking in *memorable language*
- Repeating *your core ideas* like a media brand, not a vendor

Want proof? Byron Sharp (author of "How Brands Grow") found that brands with high mental availability see exponential gains in market share—*even with low budgets*. Why? Because they're the ones people remember when it counts.

So forget about "always be closing."

Start thinking "always be recalled."

Rethink KPIs Around Visibility and Trust

Most marketing dashboards are built around lead gen metrics: CTR, bounce rate, form fills. That's fine—those still matter. But they're not enough.

Media brands measure differently. They care about:

- Reach: How many people saw it?
- Time: Did they spend time with it?
- Mentions: Are people talking about it?
- Saves/Shares: Is it valuable enough to pass on?
- Brand search: Are more people looking for you by name?

Your content should move *perception*, not just pixels.

Smarter Metrics for Smarter Content:

Metric	Why It Matters
Impressions	Top-of-funnel awareness
Post Saves	Indicates long-term utility
Direct Traffic Lift	Sign of off-platform brand recall
Brand Mentions	Tracks share of voice in-market
Average Watch Time	Proves content is holding attention

These are harder to game. But they're also more honest.

Play the Long Game of Mental Availability

Mental availability is the likelihood that someone thinks of your brand when a buying situation arises. It's built slowly, through repeated exposure to smart, relevant ideas.

This means your job is not to convert *today*. It's to become *undeniably familiar*—so that when the trigger moment hits, your brand is top of mind.

You want:

- Your visuals to be recognizable
- Your language to be distinct
- Your voice to be trustworthy

Every impression counts. Every post is a vote for what you stand for.

Action Steps

1. **Identify three high-impact ideas** from your recent content. Repackage each into three+ zero-click formats.

2. **Audit your last 10 posts.** How many required a click to get value? Rewrite three to teach in-stream.

3. **Pick two new visibility metrics.** Start tracking them in your weekly report.

4. **Create a brand recall loop.** Design a recurring content series that appears in front of your audience weekly.

5. **Brief your team.** Align everyone on the shift from click-first to impression-first thinking.

You don't need more clicks. You need more clarity. More reach. More recognition.

Because the best B2B brands aren't just clicked: They're *remembered*. They're *respected*. They're *referred*.

And that all starts long before someone ever fills out a form.

Create content that earns attention, even if it never asks for it.

Your internal media engine is what will sustain that attention.

Notes

7

The Internal Media Engine

I t's one thing to *want* to act like a media brand. It's another to build the internal systems that make it possible.

This chapter is about turning content from an ad hoc effort into a scalable engine. An engine built not on volume for volume's sake but on structure, roles, process, and rhythm.

If you want your content to work like media, you need the behind-the-scenes infrastructure that lets it run consistently, sustainably, and with increasing leverage.

Let's build that engine.

Repurposing Workflows: Blog → Video → Newsletter → Social → Sales Enablement

You do *not* need to start from scratch every time. A single strong idea can become the fuel for five different channels— if your system supports it.

Here's a high-efficiency repurposing stack:

1. **Start with a core idea.** Blog post, internal presentation, interview, field story.
2. **Pull the best quote or stat.** That's your social headline.
3. **Record a 60-second video.** Summarize the core insight.
4. **Summarize in your newsletter.** Lead with the takeaway, then link back.
5. **Reframe for sales.** Drop the key line into a one-pager or sales deck.

You're not just saving time—you're multiplying value.

Real-World Example: Encore360's "One Idea, Multiple Formats"

Every month, Encore360 builds one anchor blog post around a customer insight.

Then:

- We film a short video riff on the same topic
- We share simple quotes in graphic form on social
- A sales email is drafted using one of the takeaways
- The whole thing gets bundled into the newsletter

One idea. Multiple expressions. Maximum mileage.

Editorial Meetings as Alignment Tools

If you want consistent content, you need consistent coordination. That starts with a recurring editorial meeting.

This is not a status update. It's not a marketing stand-up. It's your newsroom.

Agenda Framework:

1. **What's coming up?** (Events, launches, timely topics)

2. **What's resonating?** (Top content from last cycle)

3. **What do we need from subject matter experts?** (Interviews, quotes, raw insights)

4. **What's getting stuck?** (Bottlenecks, delays, bandwidth)

5. **What can be reused?** (Old content ready for remix)

Invite people beyond the marketing team: sales, product, service. Cross-pollinate insights.

This meeting becomes the connective tissue. It's where ideas are born, friction is cleared, and momentum is maintained.

Internal Knowledge Capture Systems: Turn Your Team Into a Content Engine

Most B2B companies are sitting on a goldmine of expertise—and doing absolutely nothing with it.

Your sales team hears customer objections daily. Your customer service team solves the same five issues every week. Your engineers explain complex topics on internal calls that never get recorded. And your leadership team gives great off-the-cuff insights in meetings that evaporate the second the call ends.

This isn't a resource problem. It's a system problem.

If you want to build a media engine, you need a repeatable knowledge capture process—one that surfaces field-level insight and turns it into usable content fuel.

Here's how to do it:

Step 1: Run Monthly Insight Harvests

Send a recurring prompt (via email or Slack) to 5–10 trusted voices across departments:

- What's a question you've heard more than once this month?
- What's a misconception we keep correcting?
- What's a recent customer win that others could learn from?
- What's a mistake we've fixed that others should avoid?

Make it easy. Provide a Google Form or shared doc where people can drop insights in one to two lines. Or better yet— let them record a 60-second Loom.

Step 2: Tag It by Persona and Theme

Create a simple content intelligence doc with tags like:

- Persona: Engineer, CFO, Ops Manager, Regulator
- Theme: Product use, Objections, Workflow fix, New regulation

This lets your editorial team spot patterns, pitch new series, or develop Q&A-based assets instantly.

Step 3: Design a Subject Matter Expert Clip Workflow

Not everyone has time to write. That's fine.

Set up a lightweight process where subject matter experts can:

- Record async video snippets (Tella, Loom, Descript)
- Drop them into a shared drive or Notion
- Include one sentence of context (e.g., "Explaining why RH balance matters in greenhouses")

Then your team can turn it into:

- A short-form video
- A quote card
- A blog pull quote
- A new episode in a recurring series

You're not just "capturing subject matter expert insights"—you're building a living newsroom that never runs out of ideas.

Pro Tip: Add an "Insight of the Week" shoutout in your internal newsletter. Celebrate contributions. Make insight-sharing part of the culture.

Tiered Content Planning: Not All Content Is Created Equal

One reason content engines stall is because *everything is treated with equal urgency.*

A Tweet gets the same attention as a flagship article. A weekly newsletter soaks up more hours than a customer case study. Pretty soon, your team is exhausted, your calendar's bloated, and nothing feels strategic.

That's where tiered content planning comes in.

Think like a media planner. Divide your content into three clear tiers based on effort, impact, and reuse potential:

Tier 1: Fast Content (High Frequency, Low Lift)

- Social posts, carousel slides, quote cards, snippets
- Based on existing content or subject matter expert drop-ins
- Goal: Stay visible, teach something quick, reinforce POV

Cadence: 2–5x/week

Owner: Producer or content coordinator

Reuse potential: Medium (great for warm nurture)

Tier 2: Core Content (Mid Effort, High Leverage)

- Blog posts, field reports, monthly newsletters, FAQ-style pieces
- Built on audience questions or internal knowledge
- Goal: Teach deeply, build authority, feed Tier 1

Cadence: Weekly or biweekly

Owner: Content strategist or editor

Reuse potential: High (becomes social, email, sales support)

Tier 3: Hero Content (High Effort, High Impact)

- Pillar posts, white papers, video series, live webinars

- Built for major events, launches, evergreen lead gen
- Goal: Anchor your brand, generate assets that last 6–12 months

Cadence: Quarterly

Owner: Editorial lead or cross-functional project team

Reuse potential: Very high (becomes dozens of downstream assets)

This approach helps you:

- Plan realistic workloads
- Balance quick hits with long plays
- Assign the right effort to the right output

When in doubt, don't ask, "What do we publish next?"

Ask, "Which tier are we missing this week?"

Role Overlap for Small Teams: Do More With Fewer People

Not every company has a four-person editorial squad. Some have... you.

So, let's get practical. You don't need to *hire* more roles—you need to understand how roles *overlap*. And how to prioritize what matters most.

Here's how to run a media engine with just one or two people.

Minimum Viable Roles

Even the leanest team needs coverage in these four functions:

1. **Editor/Strategist** – Sets the calendar, defines what gets made

2. **Content Producer** – Creates (writes/records/edits) the content

3. **Distributor** – Posts, shares, publishes, and reports

4. **Subject Matter Expert Liaison** – Gathers insight from the field

Here's how to combine them:

If You Have...	You Need to Prioritize
1 person	Do Tier 2 content first. Repurpose for Tier 1. Drop Tier 3.
2 people	One owns content. One owns subject matter expert relationships + distribution.
Freelancer + You	You do calendar + subject matter expert work. Freelancer creates content and hands back reuse assets.

Tips for Staying Lean Without Losing Quality

- **Templates save sanity.** Create one format for blog posts, one for newsletters, one for LinkedIn posts. Reuse ruthlessly.

- **Pre-build your questions.** If you're interviewing subject matter experts, have a bank of 10 go-to prompts.

- **Use AI as scaffolding, not strategy.** AI can draft outlines, summaries, and recaps—but *you* must bring the POV.

- **Focus on *reusability*, not *quantity*.** One great blog → eight assets. That's scale.

You don't need 12 people. You need two roles covered well and a system that multiplies your work.

Style Guides and Brand Voice as Culture-Shaping Assets

Your voice is not just a tone thing. It's a *trust* thing. If every piece of content sounds different—or worse, sounds like no one—you're bleeding credibility.

To keep things cohesive and helpful, build a living style guide that defines:

- Voice and tone principles (with examples)
- Formatting rules and grammar conventions
- Common terms and how you talk about your product
- Rules for using humor, metaphors, and visuals

Make it usable. Make it real.

Then teach it. Every new hire should read it. Every contributor should reference it. And someone should own it like a product.

Pro Tip: Include a "This, Not That" section:

- We say: "practical, honest advice"
- Not: "cutting-edge, world-class solutions"

Your brand voice should help everyone write with clarity and confidence—not just the content team.

How To Keep It Lean but Consistent

You don't need a massive team. You need a clear structure. Here's a lean model that scales:

Core Roles (Can Be Fractional):

- **Editor-in-Chief** – Owns the editorial calendar and quality bar
- **Content Producer** – Writes, edits, and formats content
- **Creative Lead** – Handles design, video, and visuals
- **Subject Matter Expert Pipeline Owner** – Sources internal insights and coordinates interviews

Tools That Help:

- **Content Calendar** (Notion, Airtable, Trello)
- **Documentation Hub** (Google Docs, Notion)
- **Asset Tracker** (Spreadsheet to log repurposed pieces)
- **Feedback Loop** (Slack channel or doc for sales/customer comments)

The magic isn't in headcount—it's in clarity.

Make roles explicit. Make workflows visible. Make reuse part of the plan.

Action Steps

1. **Define your repurposing flow.** Choose one anchor format and list five ways you'll reuse it.

2. **Schedule your editorial meeting.** Weekly or biweekly, with a real agenda.

3. **Create or update your style guide.** Include voice examples, grammar rules, and brand terms.

4. **Assign internal roles.** Who owns what? Even if it's 10% of someone's time.

5. **Log your top-performing content.** Start a remix doc. Plan the next round.

Great content doesn't come from heroics. It comes from systems.

The best media teams aren't just creative. They're operational.

They plan like producers. They think like editors. They publish like clockwork.

And they treat consistency like a competitive advantage.

Build the engine once. Fuel it often. Let it run. Systems scale your business. Now let's talk about the new tools that let small teams build big presence.

Notes

8

The New Tools: AI, Video, and You

This is the chapter where things start to move faster. Technology is changing how content gets made—and who gets to make it. In the past, video production required a studio. Content writing took hours. Podcast editing was an art form. Scaling output meant scaling headcount.

Not anymore.

Now a team of two can punch like a team of ten—if they know how to use the tools.

This chapter is about how to use AI, video, and lightweight media creation to build your content smarter, faster, and with more impact—without losing your voice in the process.

Thinking Like a Showrunner in the AI Age

AI is not your replacement. It's your writing partner, your research assistant, and your intern who never sleeps.

The best content teams today treat AI like a collaborator—not a creator.

That means:

- Use it to accelerate first drafts
- Prompt it to generate structure, not substance
- Rely on it for expansion, not originality
- Edit its output ruthlessly to retain your voice

Smart AI Use Cases:

- **Outline generation:** Save time on structure, then fill it with real insight
- **Transcript summarization:** Use AI to break down long videos or webinars
- **Repurposing assistance:** Turn blogs into bullets, turn bullets into social posts
- **Headline testing:** Ask AI to rewrite a headline ten different ways, then pick the strongest

AI can give you speed, but your brain gives you *taste*. That's the part you can't outsource.

How Small Teams Can Punch Above Their Weight With the Right Tech

The gap between big-budget and no-budget has shrunk.

With the right tools, a two-person team can produce:

- Weekly video content
- Professional-looking newsletters

- Podcast episodes with full transcripts and highlights
- Consistent social media posts across multiple platforms

Essential Tools Stack for Lean Content Teams:

Function	Tool Examples
AI Writing	ChatGPT, Jasper, Claude
Video Editing	Descript, CapCut, Adobe Express
Design	Canva, Figma
Content Planning	Notion, Trello, Airtable
Newsletter	Beehiiv, ConvertKit, Substack
Transcription/Summary	Otter, Descript, Riverside

These tools don't replace your thinking. They accelerate your execution.

The trick is to use them with intention:

- Set up templates
- Reuse proven frameworks
- Batch your production cycles

Speed is a weapon—but only if it serves clarity.

Building Lightweight Video and Podcast Content

You don't need a studio. You need a webcam, a mic, and a format.

Start small. Think *repeatable*, not cinematic.

Lightweight Video Formats:

- **Talking head explainer:** One person, one insight, under two minutes
- **Screen share walk-through:** Teach one concept or feature
- **Rapid-fire Q&A:** Answer three customer questions in a row
- **Field report:** Share something you just learned or observed

Lightweight Podcast Formats:

- **Solo riff:** 5-10 minute audio take on a recent trend or question
- **Mini interview:** 15-minute expert spotlight or client convo
- **Narrated article:** Read and expand on your latest post

Batch three to four episodes at a time. Keep your production flow lean:

- Record on Zoom or Riverside
- Use Descript to cut and edit
- Republish as blog/audio/snippets

The goal isn't to impress. It's to *connect*.

Smart Prompts, Not Generic Content

Prompt engineering sounds technical. It's not. It's just asking better questions.

The better your prompts, the better your AI output. Here's how to improve your prompt game:

1. **Give context:** Tell AI who you are, what your brand sounds like, who you're writing for.

2. **Set structure:** Ask for numbered lists, outlines, frameworks.

3. **Specify tone:** Request a voice like yours—confident, clear, human.

4. **Use examples:** Feed it samples of your past writing or a style guide.

5. **Iterate:** Don't settle for the first output. Ask it to improve or rewrite sections.

Prompt Template:

"You are a B2B marketing strategist writing for [audience type]. Your tone is clear, confident, and conversational. Based on the following idea, draft an outline for a 1,000-word article that could be repurposed for LinkedIn, email, and video."

Think of AI like a sous-chef. It can chop the onions and prep the station—but *you* decide what goes on the plate.

Action Steps

1. **Choose one high-leverage tool.** Start using it this week. Set up a template.

2. **Create a repeatable video format.** Record one episode. Keep it under two minutes.

3. **Audit your content workflow.** Where can AI save you time or help you repurpose?

4. **Write a reusable prompt.** Save it for headlines, outlines, or summaries.

5. **Start small.** Don't try to automate everything. Just remove the friction that slows you down.

Operational AI Guardrails: Move Fast, Don't Break Trust

A note of caution: AI is a game-changer, but only if you use it with purpose and discipline. Without clear boundaries, it's easy to veer into generic, off-brand, or even ethically questionable territory.

That's why every content team needs a living set of AI guardrails. Not to slow you down, but to protect your voice, your credibility, and your internal trust.

Here's how to think about it:

Set a Brand-Backed AI Policy

Make it clear—internally and externally—what AI is *for* and what it's *not for*.

Your policy can be simple, but it should live somewhere people can access and update it.

Example: AI Use at [Your Company]

We use AI to:

- Generate rough outlines, headlines, and summaries
- Transcribe and condense internal interviews
- Rephrase, simplify, or restructure draft content
- Repurpose longform content into social snippets

We do *not* use AI to:

- Publish unedited output
- Fabricate quotes or testimonials
- Write first-person thought leadership
- Generate net-new articles with no human insight
- Respond to sensitive customer content

This keeps expectations clear and aligns the whole team around what "AI-powered" really means.

Red Flag Content: Human Only

Some formats *must* stay human-led. Period.

Here's your default "Human Only" list:

- Thought leadership with a personal point of view
- Customer case studies
- Anything with legal, scientific, or compliance implications
- Founder notes, origin stories, or cultural statements
- Humor, storytelling, or abstract philosophical insight

If it needs empathy, precision, or originality, don't outsource it to a pattern-matching robot.

AI ≠ Editor

AI can draft. It can riff. It can remix. But editing is strategy. It's not just proofreading—it's decision-making.

Always put a human brain between draft and publish. That's how you:

- Filter fluff
- Spot jargon
- Add nuance
- Reinforce voice
- Avoid repetition

Editing is where taste happens. And taste is how you build trust.

Pro Tip: Use Metadata

When you use AI to repurpose or summarize, tag the origin.

Example: "This summary was based on our original article, 'How Rockwool Affects Root Zone Hydration,' published May 2024."

Not only is this good practice—it helps train your team to keep editorial lineage intact, especially as reuse scales.

AI should make your work better, faster, and more scalable. But never *less you*.

Protect the voice. Preserve the nuance. Stay in the driver's seat.

You don't need more tools. You need better use of the ones already at your fingertips.

The future of content isn't just higher quality. It's higher *frequency* at higher *consistency*—without higher headcount.

Your job isn't to master every tool. It's to build a system that works *with* you.

In the AI age, creativity belongs to the teams who know how to move fast *without losing themselves*.

So, grab your mic. Open your laptop. Fire up your prompt.

Let's go make something real.

BONUS: Project Blueprint: Launch a 3-Part Video Series in One Week (With AI + Reuse)

Sometimes you don't need a content *plan*. You need a repeatable playbook you can drop in, run with, and turn into leverage by Friday.

This is that playbook.

Let's say you want to launch a short, impactful video series—something you can push to LinkedIn, email, and YouTube, while building authority in your niche.

Here's how to do it in five days, even with a small team— and with AI acting as your sous-chef, not your ghostwriter.

The Five Day Plan: "One Idea, Three Videos, Full Funnel"

Goal:

Launch a 3-part expert video series based on one core topic and turn it into multi-channel content.

Deliverables:

- 3 short videos (2–3 minutes each)
- 9 LinkedIn posts (3 per video)
- 3 newsletter blurbs
- 1 YouTube playlist
- 1 sales deck quote slide
- 1 blog round-up recap

Before You Begin: Choose Your Core Topic

Pick one idea that:

- Has buyer relevance
- Has depth (can be broken into three angles)
- Connects to your product or industry insight

Example topic:

"Three Critical Mistakes in Greenhouse HVAC Design (and How to Fix Them)"

Day 1: Plan + Outline (AI-Powered)

1. Write the *core question* your video series will answer.
    - → "What are the most common HVAC design mistakes in commercial greenhouses?"

2. Prompt AI:
    - "Break this topic into three short video outlines, each focused on one common mistake. Include a hook, three bullet points, and a closing insight."

3. Review AI output. Add your experience. Rephrase in your voice.

4. Finalize outlines for:

   ◦ Video 1: Undersized Units + Recovery Time

   ◦ Video 2: Poor Sensor Placement

   ◦ Video 3: No Integration With Climate Software

Day 2: Record (Low-Friction Setup)

1. Open Zoom or Riverside. Record one video per outline. Use bullet notes, not a full script.

2. Keep each video under three minutes.

3. Include:

   ◦ Strong hook: "If your HVAC is constantly cycling, this is probably why."

   ◦ One clear visual (diagram, screenshot, or prop)

   ◦ A single actionable takeaway

Optional: Use Descript to clean up umms, add in helpful transitions, and auto-generate captions.

Day 3: Repurpose + Caption (AI-Assisted)

1. Upload videos to Descript or Opus for auto-transcripts.

2. Prompt AI:

   ◦ "Summarize this transcript into a short social post in our voice. Include a hook and bullet points."

3. Repeat for:

    ◦ LinkedIn post

    ◦ Newsletter teaser

    ◦ YouTube description

4. Create visual quote cards with Canva or Adobe Express using punchy lines from the transcript.

Day 4: Build the Multi-Channel Kit

1. Upload all three videos to a new YouTube playlist:

    ◦ *"Greenhouse HVAC in 3 Minutes"*

2. Schedule LinkedIn posts:

    ◦ One post per video

    ◦ One quote card follow-up per video

    ◦ One "round-up" carousel on Friday

3. Drop one video per week into your newsletter as a recurring mini-series.

4. Create one sales slide per episode with takeaway quote + visual.

Day 5: Publish + Track

1. Launch first video everywhere.

2. Add teaser line to your internal Slack or MS Teams:

    ◦ *"New HVAC series just dropped—here's how to share it with leads."*

3. Use a Notion or Airtable tracker to log:

    ◦ Views

    ◦ Comments

    ◦ Internal reuse

    ◦ Sales mentions

4. Schedule your next round:

    ◦ What's the next core topic?

    ◦ Who's the next expert to feature?

    ◦ What content already exists that could spark the next three episodes?

Extra, Extra: Make It Evergreen

- Embed all three videos in a blog roundup
- Tag each with transcript + summary
- Link to it from your site's Resources section
- Repromote quarterly ("In case you missed our HVAC mini-series...")

The ROI here isn't just reach. It's rhythm.

This format builds a content cadence. It showcases expertise. It delivers utility. And it can be reused and remixed across a dozen touchpoints.

And you can do the whole thing: with one person, one subject matter expert, one AI prompt, and five days. Technology amplifies speed, but trust still compounds slowly. Let's talk about playing the long game.

Notes

9

The Long Game of Brand Authority

Most content plays for clicks. But media brands play for authority.

Authority is the compound interest of content. It's what happens when your brand consistently shows up with value, clarity, and credibility—until you're not just a source, you're *the* source.

This chapter is about how to earn that authority over time. No gimmicks. No growth hacks. Just smart, consistent execution that stacks trust until your brand becomes the publication of record for your niche.

A Few Notes on Why Clicks Are a Short-Sighted Metric

Oh, boy. Here we go. The thing your manager is not going to want to hear.

Clicks are tempting because they're easy to count. They show up in dashboards. They give teams something to report. But a click isn't a relationship. It's not trust. And it certainly isn't brand recall. It's a single moment of curiosity, often followed by nothing.

The rules of digital discovery have changed. In 2025, people skim summaries, rely on AI, and consume more content passively than ever before. In this environment, chasing clicks means playing a shallow game in a deepening pool.

Media brands understand this. They build for staying power, not spike traffic. They aim to become the brand people remember, not just the one they stumbled across once.

In the AI Era, Value Travels Without the Click

This is the biggest shift most marketers haven't fully absorbed: Your best content might never be clicked at all.

AI tools like ChatGPT, Perplexity, Arc, and Google's evolving SGE don't drive traffic in the traditional sense. They summarize. They extract. They recommend based on trust and clarity, not based on keyword tricks or curiosity-bait. In that environment, optimizing for a click is pointless. You're already downstream.

Smart media brands know this.

They design content that can stand alone in any format. A single sentence from their article might show up in an AI snippet. A stat from their newsletter might anchor a summary. The whole point is to become recognizable whether someone clicks through or not.

Clicks Are Easy to Game

Any team can drive clicks. With the right headline, a little ad spend, and some urgency, you can get people to visit a landing page. But that doesn't mean they trust you. It doesn't mean they'll come back. And it doesn't mean they'll ever think of your brand again.

Authority works differently. It's built through consistent presence, real insight, and a voice that rings true. It's not about being everywhere, it's about being remembered. Clicks don't create that. Patterns do. Patterns of showing up with something useful. Patterns of delivering value. Patterns of helping your audience think more clearly.

The best media brands don't chase attention—they earn a reputation. That reputation compounds. And it's much harder to reverse-engineer.

Why Authority Compounds

Think of authority like reputation: it builds slowly, then suddenly.

One helpful blog post doesn't make you authoritative. Neither does one slick video. Authority is the result of patterns:

- Patterns of insight

- Patterns of clarity
- Patterns of consistency

Over time, your audience begins to associate your brand with reliable answers. With useful frameworks. With ideas that get quoted in meetings.

It's not that you shouted the loudest. It's that you showed up the longest with something worth saying.

Real-World Example: OpenView's Content Engine

OpenView, a venture capital firm, became a go-to resource for SaaS operators. Not by marketing their fund, but by publishing operator-first content every week for years. Hiring playbooks. Pricing strategy guides. Growth metrics. Their blog became more useful than most industry publications. Now, when SaaS founders need guidance, OpenView is on the shortlist. That's authority, earned.

Earned Trust vs. Earned Reach

There's a difference between being widely seen and being deeply trusted. Most marketers know this intuitively, but the pressure of campaign metrics and quarterly KPIs makes it easy to forget.

Reach is an algorithmic win. It's visible. It's scalable. It makes for great graphs in a slide deck. When a post "does numbers," it feels like validation. But trust is a human win.

The brands that last aren't the ones that went viral and then rested on their laurels. They're the ones that kept showing up with content rooted in real expertise and perspectives that challenge assumptions.

It's easy to chase impressions. It's harder—and infinitely more valuable—to chase impact.

The media brands that do this are the ones who don't regurgitate what's already trending. They don't flood the zone with low-value filler just to pin down a keyword and "stay active." And they don't speak in marketing platitudes that sound like they were written by a software demo.

Instead, they show up with content rooted in lived expertise. They offer perspectives that are actually risky, ideas that poke at industry norms and make the reader stop and think. They use a voice that sounds like a human who's been in the room, not a sanitized brand trying to play it safe.

This is how trust compounds.

Not through volume, but through specificity. Not through polish, but through usefulness. Not through frequency alone, but through frequency that delivers.

Your goal isn't just to show up in the feed. It's to show up in the meeting when someone says, "I read something from [your brand] about this."

Community Building as Content Strategy

Here's the underrated truth about authority: it thrives in community.

When you invite your audience into the conversation—not just as consumers, but as contributors—you accelerate trust.

Tactics That Turn Content Into Community:

- **Feature audience questions** in your content (Q&A formats)
- **Highlight real customer wins** in your posts

- **Invite feedback** on in-progress ideas (collaborative creation)
- **Run small group webinars** or office hours with internal experts
- **Create a reader council** to preview and advise on your content

Community doesn't mean building a forum. It means treating your audience like insiders. Authority is earned by listening, then creating in public.

Becoming the Publication of Record for Your Niche

The highest goal of media-driven marketing is this: when someone wants to understand your category, they come to you first.

Not a trade magazine. Not a competitor's blog. Not Google.

You.

Think about this: Depending on your media worldview, *The New York Times* or *The Wall Street Journal* (or some similar national news outlet) fits the mold of "publication of record" in your mind. They've got the national news covered, for better or worse.

But what about your industry? What about your market segment? Which media brand is serving people who run greenhouses at research institutions? Which media brand is serving people who run rural wastewater treatment facilities? Which media brand is serving people who manufacture aerospace fasteners?

With a little bit of work, it's your business (while your competition sleepwalks into the modern digital world).

In an industry that's flooded with content, your brand needs that snap sense of memory. It needs recognition. It needs to become synonymous with clarity in a world that feels overwhelming.

The brands that are doing this well aren't worried about click-through rates as much as they are about becoming the first name that comes to mind when a question arises.

And that means thinking like a teacher, not a pitch deck.

It means showing up before the buying cycle starts; before the RFP is issued; before the prospect even knows they need you.

Because, by the time they're ready to talk, they're not looking for someone who "showed up in the feed." They're looking for the voice they already trust.

That's what media brands understand. And that's what the smartest B2B teams are learning to master.

How To Get There:

1. **Build flagship content.** Create deep, evergreen resources that anchor your category knowledge (e.g. guides, explainer videos, research).

2. **Own your beats.** Don't chase every topic. Dominate three to five key areas where your brand can go deeper than anyone else.

3. **Publish with consistency.** Weekly or biweekly content builds rhythm and recall.

4. **Establish voices.** Make your internal experts part of the brand—not just behind it.

5. **Track citations.** Are others referencing your work? That's the signal you're becoming a source.

When you do this right, you don't just participate in your market—you shape it.

The Authority Flywheel: Build Trust, Invite Feedback, Earn Status

Here we go. We love this section. Authority is earned and ultimately granted by an audience that trusts you enough to cite you, quote you, and bring you into their own thinking. That trust then drives your own content marketing momentum because you've built a feedback loop from an engaged audience.

We began to cover this in Chapter 5 when discussing how your series should function as a two-way channel. Let's go further. Here's what that flywheel looks like:

The Authority Flywheel

1. Publish With Substance

This is where it starts. Not with a campaign. Not with a keyword list. With an idea that actually says something.

To publish with substance means sharing something that's useful, original, and opinionated. Something that clarifies the conversation. Something your audience hasn't already seen a dozen times this week.

It's not enough to publish regularly or, egads, to publish with an emphasis on keywords. You have to publish meaningfully.

That means content that pulls from the field, not just internal brainstorming sessions. It means using your customer's language, not marketing-speak. It means showing (not telling) what you've learned from being in the work, solving real problems, and asking better questions.

Example: Let's say you serve facility managers. Don't just write "5 HVAC Maintenance Tips." That's filler. Instead, do a series called "The Hidden Cost of Misplaced HVAC Sensors." Show how a 12-foot misplacement added 18% to a utility bill. Include schematics. Use screenshots from real dashboards (anonymized). Pull in a quote from a customer who fixed the issue. That's a brief that someone might forward to their whole ops team.

Substance isn't just about word count or white paper polish. It's about clarity. It's about showing that you understand the problem in more depth than anyone else.

2. Earn Trust Through Usefulness

Substance is what draws people in. Usefulness is what earns their trust.

People don't remember who had the prettiest blog. They remember who helped them make a better decision. Who saved them time. Who gave them a mental model they could explain to their boss. Who helped them sound smarter in the next meeting.

This is where you start to feel traction.

A prospect DMs you after reading your post and says, *"We're sending this to our facilities director—thank you."* A customer says your article explained a problem better than their consultant did. Someone on your team references your newsletter in a QBR and the client lights up.

These moments are your proof-of-usefulness. We keep reiterating these because they are important qualitative metrics. They tell you your content is working—not just as marketing, but as an operational resource.

This is what real thought leadership feels like. Not vanity metrics. Not fake authority. Just honest, helpful, human guidance that your audience keeps turning to.

You can already hear the flywheel turning...

3. Invite Participation

Once you've earned some trust, don't let it stagnate. Put it to work.

Invite your audience *into* the process. Give them a stake in your series. Make it feel less like a brand broadcast and more like a standing conversation.

Ask them what they're stuck on. Use simple, specific prompts:

- "What's the most annoying part of spec'ing a new HVAC system?"
- "What's one thing you wish regulators understood about your job?"
- "What's your boldest prediction for the next 12 months?"

Add these as CTAs in your newsletter. Drop them into your LinkedIn posts. Include a note in your podcast description: "Got a question for our next guest? DM us."

When you get responses, feature them. Quote your audience. Share their pain points. Turn their language into your headlines.

Example: At the end of your monthly "Field Guide" newsletter, include: "What's the biggest compliance headache you've faced this year? Reply and we'll feature a few in next month's Field Guide."

This is good editorial research. And it makes your audience feel seen, not targeted.

4. Create With the Community

Now we're really cooking.

When you use the questions and insights from your audience as raw material, you're no longer just creating for them—you're creating *with* them.

Build a column called "You Asked." Film a quick series of two minute video answers. Invite customers to submit a question and provide a public reply. Write a blog that credits three commenters from a recent LinkedIn post.

You don't need to be fancy. You just need to be consistent—and generous with attribution.

This kind of co-creation deepens your authority because it shifts your brand from a voice about the industry to a voice inside it. You stop sounding like a vendor trying to "speak the language" and start sounding like the people your audience actually trusts: their peers.

It also keeps your content pipeline full of real problems, real language, and real stakes. You don't have to guess what to write about next. Your readers are telling you.

And the more your audience sees themselves in your content, the more they'll advocate for it. Internally. Externally. Consistently.

5. Earn Citations, Mentions, and Loyalty

This is the payoff. And it's not a flashy one, it's a cumulative one.

Your content starts to show up in places you don't control. In Slack threads. In meeting decks. In internal knowledge bases. In sales calls where a prospect references your post before your AE brings it up.

This is the moment you graduate from content producer to trusted voice. You're no longer just publishing. You're shaping the conversation. You're influencing the language people use to talk about the work.

Sales tells you that a prospect came in warm because they "already feel like they know us." A customer says your newsletter is the only one they read. Your podcast starts getting quoted in trade publications.

You're not chasing virality. You're earning embeddedness.

This is when the flywheel really starts spinning. Every piece you publish has more lift because the groundwork has already been laid. Your audience knows your voice. They trust your intent. They look forward to what's next.

That's authority. Built slowly. Earned honestly. Powered by repetition and relevance.

And here's the kicker: now, when you publish the next post, it doesn't feel like the beginning of a new campaign.

It feels like the next chapter in a conversation they already care about.

Why This Works

The authority flywheel closes the gap between publishing and listening. Now you're acting like a media brand (more on this in the next chapter).

Too many brands still operate like megaphones. Authority brands operate like conversations.

The more your audience sees themselves in your content, the more likely they are to reference you when it matters.

Long-Term Authority Calendar: Your Editorial Rhythm for Earning Industry Gravity

Authority doesn't spike. It stacks.

Which means you need a calendar not built for clicks— but built for trust. Not just content cadence—but reputation architecture.

Here's a quarterly content calendar designed for compounding authority over time.

Quarterly Authority Calendar Blueprint

Every quarter, repeat this structure:

Week	Focus	Output
Week 1	**Flagship Drop**	One in-depth guide, report, or explainer. Built to last. Optimized for citation.
Week 2-3	**Deep Dive Follow-ups**	Two to three social posts, carousels, or short videos expanding on the flagship's key ideas. Reuse quotes, states, visuals.
Week 4	**Community Activation**	Host a small webinar, office hours, or AMA related to your flagship. Collect new questions and quotes.
Week 5	**Collaborative Content**	Publish a round-up or "Voices From the Field" post using input from readers or customers.
Week 6	**Curation + Reflection**	Recap what's been learned. Pull it together in a newsletter or blog. Reinforce your POV.

Repeat with a new "beat" or theme next quarter.

Annual Authority Stack: Go Big on Structure

At a macro level, think of quarters as themes and months as content tempo.

- **Q1:** Education and Planning
  - Drop your "202X Outlook" or Strategic Planning Toolkit

- **Q2:** Problem Solving
  - Drop a big "State of the Industry" or "What's Broken and How To Fix It" anchor piece

- **Q3:** Optimization and Tactics
  - ◦ Run a hands-on series: guides, templates, operator interviews
- **Q4:** Vision and Review
  - ◦ Drop a "Year in Review," future trends, or key takeaways from a survey or event

Each quarter should have *one* major tentpole asset that you design everything else around.

Metrics to Track

Authority is squishy—but here's what you *can* measure:

Signal	Metric
Influence	Direct traffic lift, branded search growth
Community	Email replies, LinkedIn post comments, webinar Q's
Utility	Content reuse by internal teams
Citability	Backlinks, shares, GPT citations (!), being referenced in buyer convos
Presence	"We saw your piece on..." mentions in calls or sales notes

Build To Be Quoted

If your content isn't getting bookmarked, sent to someone's boss, or cited in a planning meeting... it's not yet authoritative.

But if you commit to a rhythm—one tentpole idea, one monthly series, one community touchpoint—you don't just keep the engine running.

You start building something no one else can replicate:

A content brand that sets the tone for your entire category.

Action Steps

1. **Audit your content for authority signals.** How many pieces are definitive, deep, or unique?

2. **Pick one beat to own. Go deep.** Publish three to five high-value assets in the next quarter.

3. **Invite your audience in.** Ask for their questions, highlight their wins, build with them.

4. **Elevate a voice.** Choose one subject matter expert or exec and build a content series around their POV.

5. **Design a publication mindset.** Build an editorial calendar like a media company, not a campaign machine.

About That Authority Audit... Here's How To Score What You've Built, Then Build What You're Missing

Let's be honest: Most brands *think* their content is authoritative.

It's well-written. It's informative. It checks the SEO boxes.

But, when you look closer, it's often surface-level, reactive, or indistinct.

That's where an authority audit comes in.

Think of this as your editorial quality check. Not for grammar or polish—for *influence*. For credibility. For the kind of content that earns citations, shares, and trust.

Here's how to do it:

The Five Signals of Authority Content

Score each content asset from 1–5 across the following dimensions:

Signal	Question to Ask	Score Criteria
Depth	Does this go beyond the obvious?	1 = skimmable fluff, 5 = expert-level breakdown
Originality	Are we saying something *only we* could say?	1 = generic rehash, 5 = proprietary POV or framework
Utility	Can the reader do something with this immediately?	1 = vague, 5 = immediately actionable
Citability	Are there moments that are quote-worthy or reusable?	1 = forgettable, 5 = "Wait, I'm saving this"
Consistency	Is this aligned with our known voice and beats?	1 = off-brand or one-off, 5 = totally on-message

Then calculate your average. Anything scoring under a three? Put it on a list for refresh, upgrade, or retirement.

Bonus Layer: Add Metadata Tags

While auditing, tag each piece for:

- **Persona relevance** (e.g., CFO, Engineer, VP Sales)
- **Funnel stage** (early education vs. late evaluation)
- **Repurpose potential** (could this be broken into snippets, slides, or quotes?)

This not only cleans up your library—it also makes future planning exponentially easier.

Pro Tip: When in doubt, ask: *Would a customer reference this in a board meeting?*

If yes—it's authoritative. If not—it might just be content.

You don't earn authority by sounding important. You earn it by being indispensable.

Useful beats loud. Consistent beats clever. Trusted beats trendy.

Your job isn't just to say something smart. It's to become the brand people trust to make *sense* of what's happening.

Because when your content becomes the compass for your audience, you don't have to chase leads. They already know where to find you.

Authority is the endgame. But you have to step fully into your role as a media company to earn it.

10

Your Company Is a Media Company (Whether You Know It or Not)

This is the final chapter. And it's time to say the quiet part out loud:

If you're doing business in 2025, you are a media company. Full stop.

You may sell manufacturing equipment, or enterprise software, or consulting services, but the moment you started publishing content online, you entered the media arena.

The only question now is: Are you acting like a brand that *owns* its message or one that's just reacting to the market?

Let's talk about what it means to step into this role with purpose.

Making the Mindset Shift From Vendor to Voice

Most companies still think like vendors: They show up when it's time to pitch. They talk about themselves. They publish content that reads like a brochure. And then they wonder why nobody's listening.

Media brands do the opposite. They:

- Publish regularly
- Prioritize audience value
- Build personality and point of view
- Show up even when there's nothing to sell

The goal is no longer to "get in front of people." The goal is to *become* the brand people already pay attention to.

That means:

- Investing in signature content formats
- Building trust before the sales process starts
- Being known for something beyond your product

And it's not optional. It's table stakes.

Real-World Shift: From Vendor Voice to Media Voice

Old way: "Our product helps optimize workflows with scalable AI." Media way: "Here's what your ops team is really struggling with—and what top performers are doing differently."

Same expertise. Different energy. One pushes, the other attracts.

How Your Brand Becomes a Destination

The most valuable B2B brands are not just product pro-viders. They're *learning environments*.

They publish guides, host webinars, run newsletters, release data. They don't just sell solutions—they set the agenda.

When you become a destination, your brand becomes:

- A source of clarity
- A shortcut to insight
- A trusted companion through change

People don't just tolerate your content. They *seek it out*.

How to build that:

- Define your editorial focus: What does your brand cover better than anyone else?
- Commit to frequency: Weekly or biweekly content builds audience memory.
- Develop characters: Founders, subject matter experts, voices your audience knows by name.
- Build feedback loops: Let your audience guide what you publish next.

You don't need a media empire. You need a media *posture*.

A Checklist for Launching Your Own B2B Media Flywheel

Want to build like a media brand? Here's your starter kit:

1. **Clarify Your POV**

    ◦ What does your brand believe? What's your take on the industry? Be clear. Be bold. Be useful.

2. **Launch a Flagship Format**

    ◦ Choose one high-impact format: a newsletter, a show, a series. Name it. Promote it. Build it like a product.

3. **Layer Distribution Channels**

    ◦ Map your rented (social), owned (site/email), and internal (team voices) distribution system. Build a weekly rhythm.

4. **Activate Your Experts**

    ◦ Don't just quote your subject matter experts— turn them into contributors. Ghostwrite. Film. Interview. Promote their brains.

5. **Build a Repurposing System**

    ◦ Every core idea should live in multiple formats. One anchor, five expressions.

6. **Track Smart Metrics**

    ◦ Impressions. Shares. Saves. Mentions. Brand search volume. These tell you if your content is working beyond clicks.

7. **Operationalize Your Editorial**

- ◦ Have an editorial owner. Hold regular planning meetings. Use a real calendar. Treat it like production.

8. Prioritize Consistency

- ◦ The brands that win aren't always the ones with the flashiest content. They're the ones who show up—every week, without fail.

Keep those steps top of mind in editorial meetings. That's how you should reframe your marketing meetings now: as editorial meetings.

Start doing this now to reap the benefits of habits later.

We opened this book with a sense of urgency about how to think about marketing and media. The world is changing fast. Let's return to that theme by zooming out and addressing the broader trends in media distribution head-on.

Distribution Has Changed. So Must You.

Let's end where most marketers begin: with distribution.

For decades, distribution was treated like logistics. Get the message. Pick the channel. Push the button. Done.

That model worked when the pipes were predictable. When audiences still clicked. When SEO wasn't a bloodsport. When "reach" felt like something you could buy.

But in 2025, the mechanics of distribution have shifted under our feet.

Let's say it plainly: You no longer distribute content. You distribute identity.

That's the inflection point we're living through. If you don't see it yet, you're already behind.

The Myth of the Open Channel

Marketers used to treat the internet like a highway. You published something, put it in the right lane—email, LinkedIn, Google—and assumed it would reach your audience. It might slow down in traffic, but it would get there.

That's no longer the case.

In 2025, *everything is throttled*. Organic reach on social is artificially capped. Email is triaged by AI. Search is becoming answer, not navigation. And user behavior is fragmented across dozens of screens, apps, formats, and feeds.

The pipes are still there—but the water pressure is gone.

There is no longer a highway. There are alleyways, shortcuts, secret doors, and personal invitations.

Distribution now runs on trust, velocity, and identity. And unless you've built those, your content doesn't go anywhere.

The New Rules of Distribution

So what does *work*?

1. You're not an account. You're a voice.

People follow people. Not brands. Not logos.

And people follow not because they clicked a CTA but because they heard something true, clear, and useful.

Your distribution strategy is only as strong as your voice.

2. You earn distribution through memory.

Frequency isn't spam. It's a trust-building mechanism.

If your content shows up with value every week, your audience starts to recognize it.

That recognition is what makes them stop scrolling next time.

3. You distribute through resonance, not reach.

One quotable paragraph in the right executive Slack channel is more valuable than 5,000 impressions.

One chart saved and reused in a board meeting is more powerful than an entire email drip.

4. You don't win the feed. You win the follow-up.

In a world where no one clicks, your job isn't to drive traffic.

It's to plant ideas. Ideas that re-emerge in conversation.

"Where did I see that post about root zone balance?"

"I think that newsletter said it better."

That's the moment your brand wins.

What Distribution Really Means Now

Distribution used to be about placement. Now it's about presence.

And presence can't be automated. It can't be faked. It can't be bought. Presence is earned—through rhythm, voice, clarity, and repetition.

You don't distribute a blog post. You distribute a perspective. You distribute a personality. You distribute a body of work that is coherent, valuable, and unmistakably yours.

And That Brings Us to the Real Job of the Modern B2B Marketer:

Not to chase algorithms. Not to flood the feed. Not to reverse-engineer "what works."

But to show up like a source.

To know your beats. To know your reader. To write and publish with the same calm, steady rigor that the best media brands always have.

Because the future of distribution is not a trick. It's not a tool. It's not a platform update.

It's a return to craft. It's what media companies have always understood: Make something *so useful, so resonant, so clearly from you*—that people want more of it.

That's how you move through the noise. That's how you build a brand that travels without a push. That's how your content shows up in meetings you're not in, search results you didn't pay for, and decision-making cycles that started long before your SDR picked up the phone.

That's the distribution engine you're building now.

And that's what it means to be a media brand in 2025.

Final Challenge: Think Like a Showrunner, Act Like a Publisher, Serve Like a Teacher

This book has been about one big idea: B2B marketing teams can—and should—act like media brands.

That doesn't mean becoming BuzzFeed. It means:

- Thinking like a showrunner: What stories are we telling? Who are the characters? What's the format?
- Acting like a publisher: What's our cadence? Who's our audience? What's our voice?
- Serving like a teacher: What do our readers need to understand, overcome, or achieve?

You don't need more hacks. You need a new posture.

One built on clarity. One built on service. One that earns trust before it asks for attention.

Because when you build like a media brand, you don't just fill the funnel—you *become the source*.

Now go build something people want to subscribe to. Your audience is already listening.

You are the media now.

Make it count.

Notes

Epilogue:

The Shift Is Yours To Make

By now, you've seen the pattern.

The future doesn't belong to brands with the biggest budget. It belongs to those with the clearest voice. The most consistent cadence. The strongest point of view.

It belongs to companies willing to stop acting like vendors and start behaving like media.

This isn't a trend; it's a shift. And it's already underway.

You've felt it, haven't you? The fatigue of trying to out-shout the market. The endless scramble for clicks. The content treadmill with no clear payoff.

You've also felt the thrill. The resonance when something you publish *lands*. The email reply that says, "This is exactly what I needed." The moment your brand isn't just a product—but a presence.

That's the difference between marketing that interrupts and marketing that *matters*.

And here's the truth: You don't need permission. You don't need perfection. You don't even need a massive team.

You just need the guts to show up with something real, again and again.

So here's your invitation:

Be the voice of clarity in your market. Be the source your audience bookmarks. Be the brand that helps people think sharper, act faster, and feel like they're not alone.

That's the job now. Not to chase trends. Not to spam channels. But to build trust, in public, at scale.

To teach. To serve. To lead.

Because the best brands aren't focused solely on selling. They're publishing. They're documenting. They're building movements.

The media mindset is a long-term posture. One that doesn't fade with algorithms.

So go make your corner of the internet smarter. Start the series. Post the idea. Share the insight.

Not for clicks. Not for vanity. But because your audience deserves better—and you're one of the few willing to give it to them.

We'll be out here with you. Publishing. Listening. Learning.

Let's make something worth following.

The shift is yours to make. And it starts now.

Appendix:

Making It Real

Part 1: Handling Common Objections (Convince Your Boss Kit)

Objection 1: "We don't have time for content."

Response: You don't have time to keep chasing cold leads that don't convert. Content saves time by warming prospects before sales ever picks up the phone. Start small. One series. One subject matter expert. One month of consistency. Prove it, then scale.

Objection 2: "Our buyers don't read content."

Response: Your buyers read *great* content. They listen to niche podcasts. They forward smart newsletters. They quote sharp LinkedIn posts in meetings. The issue isn't content—it's quality. The right content, delivered in the right format, works.

Objection 3: "We don't have anything interesting to say."

Response: If your team is solving real problems, you have stories to tell. Start with customer questions, internal expertise, or what your reps hear daily. Turn problems into media. Use the voice of the operator, not the marketer.

Objection 4: "We can't afford to build a media team."

Response: You don't need one. This book was written for *small teams*. The tools are lean. The formats are light. Start with one anchor and a few repurposed outputs. Content doesn't require scale—it requires consistency.

Objection 5: "How will this drive leads?"

Response: Trust drives leads. Content builds trust. When your brand is consistently useful, top of mind, and easy to engage with, it reduces friction across the funnel. And when someone *is* ready to talk, they call you—not your competitor.

Part 2: Beat-To-Buyer Strategy Map

Buyer Type	What They Care About	Content Style	Format Examples
Decision-Maker	Strategy, ROI, long-term impact	Sharp, authoritative	POV blogs, newsletters, trend briefings
Economic Buyer	Cost, value, de-risking investment	Clear, confident	ROI calculators, buyer guides, exec memos
End User	Ease of use, pain points, solutions	Practical, conversational	How-to posts, short videos, playbooks
Technical Gatekeeper	Compatibility, performance	Detailed, data driven	Tech specs, case studies, integration guides
Influencer	Credibility, innovation, brand rep	Insightful shareable	Podcasts, social posts, expert interviews

Pro Tip: Use this table to audit your next 10 pieces. Are you covering each role? Are your formats aligned with their mindset?

Part 3: Your First 30 Days – A Media Mindset Launch Plan

Week 1 – Kickoff + Core Assets

- Define your point of view: What does your brand believe?
- Pick two to three core content beats (problems you solve)
- Identify one subject matter expert to partner with
- Choose one flagship format (newsletter, blog, video series)

Week 2 – Build Your Systems

- Set up a basic editorial calendar
- Write your first three content pieces
- Create a repurposing template
- Draft your brand voice guide (with a few "This, Not That" examples)

Week 3 – Publish + Distribute

- Launch your first piece
- Share it across 3+ channels (owned, rented, internal)
- Ask your subject matter expert to share it with their network
- Capture early feedback (engagement, questions, shares)

Week 4 – Evaluate + Scale

Review what landed: What got attention? What was ignored?

Repurpose your best-performing piece into three formats

Hold your first editorial sync with two cross-functional reps

Plan your next four weeks of content using the flywheel model

By Day 30, You Should Have:

- A clear content focus
- One published series in motion
- A repurposing engine running
- Internal alignment around voice and cadence
- Enough momentum to keep going

This isn't about perfection. It's about rhythm.

The brands that win don't wait until the strategy is flawless. They ship early. They listen hard. They get sharper with every post.

So take the next step. You're not starting from scratch. You're starting from clarity.

Build the flywheel. Run the play. Earn the trust.

The next 30 days could be the beginning of a whole new way your market sees you.

Part 4: Budget and Resource Realism

How to Run a Media-Driven Marketing Strategy Without a Studio, an Agency, or a Six-Figure Budget

Let's kill a myth real quick: you don't need a full-scale content team to execute this playbook. You don't need a 12-month production schedule, a podcast studio, or a fleet of videographers. You just need a system—and a little clarity about what *actually* moves the needle.

This appendix is your sanity check. A field-tested resource for marketing teams with one, two, or three people trying to figure out: "What can we realistically do, and how do we make it work without burning out?"

The One-Person Media Company

If it's just you, here's your play:

Priority: Consistency over volume.

Strategy: Pick one format and one distribution channel. Build rhythm before reach.

You are: The strategist, the writer, the editor, the distributor, and the analytics team. That's okay—for now.

Your Minimum Viable Stack:

- **Newsletter platform:** Beehiiv, ConvertKit, or Substack
- **AI assistant:** ChatGPT or Claude for outlines, rewrites, and repurposing
- **Design:** Canva templates for quote cards or episode covers

- **Content tracker:** Notion, Trello, or a spreadsheet—whatever you'll actually use
- **Video (optional):** Descript or CapCut for short lo-fi clips

Your Weekly Cadence:

- One anchor content piece (blog, newsletter, or video)
- Three repurposed social snippets
- One remix post (reuse something older, freshened with a new hook)

You're not trying to boil the ocean. You're building habits, credibility, and muscle memory.

The Two-Person Marketing Team

If you've got a partner, you've got a system.

Priority: Division of labor + speed of execution

Strategy: One person drives production. One person drives strategy, planning, and distribution. Alternate weekly if needed.

You are: A tag team. Not a content mill.

What Changes:

- You can add a second format (e.g., blog *and* video, or podcast *and* newsletter)
- You can batch-create content 2–3 weeks out
- You can schedule, analyze, and adjust without losing momentum

Suggested Roles:

- **Producer:** Owns content creation (writing, editing, filming)
- **Publisher:** Owns calendar, distribution, repurposing, and analytics

Tool Upgrades:

- Shared Notion calendar with recurring series
- Slack or email summaries of content performance
- Canva Pro for brand kit consistency

Your job now isn't to double output—it's to double *quality per hour*. Let the system do the heavy lifting.

The Three-Person Team

With three, you have an engine. Now it's time to scale smart.

Priority: Editorial ops and internal knowledge capture

Strategy: Treat content like a product. Add a true content owner and bring in subject matter experts regularly.

Suggested Roles:

- **Editor-in-Chief:** Owns narrative, themes, and calendar
- **Content Producer:** Handles production of posts, emails, and creative assets
- **Distribution Lead:** Manages publishing, repurposing, subject matter expert coordination, and audience growth

New Capabilities:

- You can launch a recurring video or podcast series
- You can build and test lead magnets
- You can systematize subject matter expert involvement without burning them out

Tool Additions:

- Riverside or Zoom + Descript for video/audio capture
- Figma or Canva + Loom for visual storytelling
- Airtable or Trello for editorial pipeline visibility

This is where you stop sprinting and start building scale. Your goal is leverage—making each idea go further, show up in more places, and move the right people closer.

How to Prioritize When You Can't Do It All

1. **Pick one channel where your audience *already* spends time.** Don't chase every platform.

2. **Commit to one flagship format.** Newsletter, video series, blog, podcast—choose one and get great at it.

3. **Build once, publish five ways.** Every piece should have a clear remix plan (see Chapter 7).

4. **Create a "What To Skip" list.** Kill the busywork: vanity reports, over-designed PDFs, one-time cam-

paigns that won't age well.

5. **Revisit every 90 days.** Look at what's landing. Sunset what's not. Keep it nimble.

Final Word: Constraints Are Creative Fuel

The best media teams aren't the biggest. They're the ones that plan like showrunners, think like journalists, and publish like pros—regardless of headcount.

If you're lean, good. It'll keep your instincts sharp.

If you're scrappy, even better. It'll force clarity. If you're understaffed, no problem. You're reading the right book.

Just build the engine you can actually run. That's the important thing that media brands do every day: They ship good content to their audience. And as the days stack up, they build trust. That trust becomes the bedrock of their business.

Part 5: An Interview With Encore360 Content Director Eric Sandy

Q: What's the origin story of *The Media Mindset*?

A: Honestly, it came out of frustration. We kept seeing the same pattern: good marketing teams with good instincts stuck in short-term cycles. One-off campaigns. Scattered blog posts--and that's when the blog would get updated once every six months. That sort of stuff leads to burnout, and it's just bad for business. I had a background in journalism, and I knew there was another way: content as infrastructure, like a news outlet. We started building around that idea, tested it across industries with our clients, and it worked. The book was the natural outcome. We also just like books here at Encore.

Q: You say marketing isn't here to sell—it's here to build trust. What does that look like day-to-day?

A: It means showing up even when you don't have something to promote. Maybe *especially* when you don't have something to promote. Just stop with the product promotion in every newsletter. It means writing what the audience needs, not what your internal teams want to shout. That can cause some internal tension, by the way. It's not intuitive for a lot of businesses. If you're doing it right, sales becomes the second conversation, not the first.

Q: B2B isn't boring, it's under-produced. What's a good example of that?

A: The water and wastewater industry. Most people think it's dry, technical, too niche. But once you start talking to operators and engineers, you hear wild stories—chaotic repairs, regulatory stress, institutional knowledge that lives in one person's head. That's great content. It just needs a format and some respect. That's partly why we launched our own media brand, *Water Daily*, to just get out in the market and start covering news features. It's been super fun and it helps us prove the concept to our clients.

Q: Why is campaign-mode such a trap?

A: Because it resets your momentum every time. You get a spike of effort, some clicks, maybe a few leads—then you disappear. It takes a lot of energy each time and there's no memory being built. Thinking in series solves that. You start to earn audience recognition. You stop asking, "What should we publish?" and start asking, "What's the next episode?" This is one of the big messages in the book, of course.

Q: What's one media habit you wish every B2B team would adopt?

A: An editorial meeting. Not a status check; a real editorial conversation. Bring in sales. Bring in customer success. Talk about what people are asking in the field. That's how you make content that matters. Scrap all your marketing meetings and turn them into editorial meetings. Even just using that word will change your mindset.

Q: How do you respond to "zero-click content isn't converting"?

A: I'd ask what we're optimizing for. If you want people to trust you, start by being useful, even if they never click.

The best leads I've seen came from people who had been reading for months without engaging. Zero-click content builds memory. And memory builds inbound. It's hard to track, sure, but it's the way of the world. I wish it wasn't, but I also wish I was good at golf. Some things are just beyond our control.

Q: "If your content doesn't get reused, it didn't hit." That's from the book. Can you unpack that?

A: It's about utility. If no one's quoting it, forwarding it, dropping it into a deck, it probably didn't resonate. When content works, it travels. Internally. Externally. Reuse is the clearest sign that you've earned attention and built clarity. Think about things you've shared lately; maybe it was about your favorite baseball team or a local news story. It's the same impulse.

Q: What's your take on AI in all of this?

A: I'm a reluctant convert to using AI tools all day long. Some of these tools remove the friction that old search habits had ingrained. Some of them allow me to talk through problems and receive very targeted solutions, and it's not always about work. Could be about my golf swing. AI is great at assessment, synthesis, and iteration. Those are important modes for getting stuff done in your life. It's terrible at taste. So, yes, I use it all the time as a content guy—for assessing a given piece of content, for synthesizing disparate ideas, for iterating on something that I think is good but maybe not quite there. But I still trust my gut on what's actually good. If AI helps me move faster without sounding like I outsourced my brain, I'm in. That's the trick, too: You can't let it think for you. I could go on, but that's a real societal risk. Don't let your marketing team fall into that trap.

Q: What's your advice for the two-person team trying to act like a media brand?

A: Pick one series and one channel. We went hard on LinkedIn as a content business. Build rhythm, cover the news. Forget everything else until that's consistent. You're not trying to be everywhere. You're trying to be memorable somewhere. As of the writing of this book, Encore360 is a three-person team. We're growing, and we'll be a larger team by the time you read this, but I guess my point is that it can be done with a small team.

Q: What's the biggest content myth you'd love to kill?

A: That volume equals value. I'd rather see one strong series run weekly than 14 blog posts that no one remembers. Frequency matters, but clarity comes first. I was a longform magazine writer for many years and I love the format, but the 2,000-word blog post that your team converts to a 3,000-word e-book just doesn't hit in a marketing sense. Stop that. Leave the long stuff for The New Yorker.

Q: What's one battle you've had to fight to keep content sharp?

A: Saying no to C-suite pressure. Once a marketing team starts delivering, everyone is suddenly a writer or a designer. But that's not true. Your chief operating officer shouldn't be reviewing your content and adjusting your voice. But I see this all the time.

Q: What non-marketing thing teaches you the most about content?

A: Hiking. There's something about being a few miles into the woods with no service that forces you to think clearly. Every good idea I've had started with silence. It teaches you pacing, too, which is basically what editorial rhythm is. Fresh air solves most short-term problems.

Q: What's a piece of media you remember exactly where you were when you encountered it?

A: That's a tough one. Of course, there are so many. One might be another book, "About a Mountain" by John D'Agata. It's the only book I've read in one sitting, in our old apartment in downtown Cleveland. The book is paired with another that D'Agata co-wrote with his editor, and *that* book is all about how he wrote the first book, about challenging the basic facts that D'Agata wrote (and sometimes contorted). Both books really blew my mind.

Q: If your content philosophy had a soundtrack, what's track one?

A: "Once in a Lifetime" by Talking Heads. Frankly, anything by Talking Heads. This book makes it seem like the world is changing and everything is new, but the core truth is that everything we've written here is the same as it ever was. There's nothing new in this book. But we believe that it's very easy for people to forget core truths. That's why it's so important to reiterate them in new ways. You don't stop telling your spouse you love them just because it's been said, right?

Q: What's your current media diet?

A: Oh, the usual. A mix of B2B and industry-specific news-letters in the morning, The New York Times and The New Yorker for national stuff. Longform journalism (RIP longform.org) and good sportswriting. Anything that respects my time but still surprises me. And, well, maybe to contradict that last part about respecting my time-- I read a lot of reddit.

Q: What's something B2B marketers could learn from late-night TV or stand-up comedy?

A: Segment discipline. Every great show has a structure: monologue, desk bit, interview, closer. (Just like news broad-casts, too.) They know how to open strong, land a punchline, and keep the rhythm. If your content series can do that, you're already ahead. And try to leave your audience a little bit happier than they were before they found you. Try basing a short B2B video off the structure of an SNL sketch. You might just delight yourself.

About the Authors

Jim Gilbride

With over 20 years of experience in the media industry, Jim Gilbride has a proven track record of building and transforming brands to drive audience engagement and deliver client success. His expertise in media, marketing, and operational strategy has helped grow loyal audiences and create innovative marketing solutions that generate measurable ROI. From revitalizing underperforming brands to launching new ventures, Jim's leadership has consistently connected clients with engaged, high-value audiences.

Scott Anthony

With over 12 years of experience in media and content marketing, Scott Anthony is an accomplished leader in driving client success through collaboration. He takes a needs-based approach, investing time to fully understand each client's business and objectives before recommending solutions. Scott's ability to align tailored strategies with specific business goals ensures measurable results and long-term value for client success.

Eric Sandy

Eric Sandy is an award-winning editorial leader with over 18 years of journalism experience, specializing in B2B marketplaces and driving digital audience engagement. In his previous role, Eric successfully grew a B2B media website from 8,000 to over 200,000 users in a highly competitive environment by identifying editorial gaps and delivering original, impactful content. His vision for content creation and deep understanding of niche audiences fuel the Encore360 mission.

Notes

Notes

Notes

Notes

Notes

Notes

Notes

Notes

Notes

Notes

www.ingramcontent.com/pod-product-compliance
Lightning Source LLC
Chambersburg PA
CBHW021929190326
41519CB00009B/956